AMERICAN VACUUM SOCIETY
CLASSICS

FIELD EMISSION AND FIELD IONIZATION

FIELD EMISSION AND FIELD IONIZATION

Robert Gomer

The University of Chicago

Library of Congress Cataloging-in-Publication Data

Gomer, R. (Robert)
   Field emission and field ionization/Robert Gomer.
      p. cm.—(AVS classics of vacuum science and technology)
   Originally published: Cambridge: Harvard University Press, 1961.
   Includes bibliographical references and index.
   ISBN 978-1-56396-124-3
   1. Field emission. 2. Field ion microscopy. I. Title II. Series.
QC700.G66   1993
537.5'3–dc20
                                             92-46417
                                             CIP

To Dicky and Maria

Series Preface

The science of producing and measuring high and ultrahigh vacuum environments has fundamental interest for basic research in addition to a wide variety of important practical applications. Basic research involving particle physics, atomic and molecular physics, plasma physics, physical chemistry, and surface science often involves careful production, control, and measurement of a vacuum environment in order to perform experiments. Practical applications of vacuum science and technology are found in many materials processing techniques used for microelectronic, photonic, and magnetic materials and in the simulation of space and rarefied gas environments.

As in most of modern science, the rapid development of the field has been accompanied by a parallel growth in the related technical literature including specialized journals, monographs, and textbooks. There exist early publications in vacuum science and technology which have attained the status of indispensable references among practitioners, lecturers, and students of the field. Many of these "classic" publications have gone out of print and are currently unavailable to newcomers to the field. The present series, commissioned by the American Vacuum Society, and published within the Book Program of the American Institute of Physics to celebrate the 40th anniversary of the Society in 1993, is entitled the "American Vacuum Society Classics."

The American Vacuum Society Classics will reprint important books from the last four decades that continue to have significant impact on the modern development of the field. It is the goal of the American Vacuum Society Classics to reprint these books in a high quality and affordable format to ensure wide availability to the technical community, individual researchers, and students.

H. F. Dylla
Continuous Electron Beam
Accelerator Facility
Newport News, VA
and
College of William and Mary
Williamsburg, VA

Author's Preface to the Reprint Edition

Some thirty-one years have passed since the publication of *Field Emission and Field Ionization* by the Harvard University Press; my children to whom the book was dedicated are themselves parents. When this book was written field emission was one of the very few techniques available for surface studies and the attainment of ultrahigh vacuum was an art known only to a handful of practitioners, rather than something to be bought off the shelf. It is gratifying that there should still be enough interest in field emission and ionization to warrant a reissue, but clearly time has not stood still even in these areas. This preface attempts to provide at least a minimum of necessary updating and corrections. By and large the descriptions of both field emission and field ionization in this book are still valid, although more sophisticated treatments of field emission[1] and of energy distributions, particularly in the presence of adsorbates,[2] and of field desorption[3] have since been given. There is only one outright error that I am aware of: the derivation of the enhancement factor for gas supply, Eq. (16), p. 73 assumed real velocity at $r=r_t$ rather than at all $r \geqslant r_t$. Southon[4] showed that the correct enhancement factor is $\sim(\pi \alpha F^2/2kT)^{1/2}$.

What about applications? Field emission is no longer at the forefront of most chemisorption or physisorption research and is used only occasionally for measuring work function changes caused by adsorption. In certain cases, mainly on the $W(110)$ plane for some but not all adsorbates, the segmentation of the Fowler–Nordheim equation, into field dependent and independent parts breaks down and work function changes are more reliably determined via Kelvin probe techniques.[5] Field emission energy distributions in the presence of adsorbates provide information on local density of states with great surface sensitivity, but the rapid decrease in emission with decreasing energy limits the method to 0–2 eV below the Fermi level, unless very special analyzers are used.[6] This method has effectively been supplanted by photoelectron spectroscopy. The most useful application of field emission to surface science is still the study of surface diffusion, although no longer via the shadowing method described in the book. D can be measured on single planes at constant coverage (which can be varied) by the current fluctuation auto-correlation function method.[7] This consists of covering an emitter uniformly with adsorbate and measuring the current from a small region of the surface via a probehole. When there is random diffusion of adsorbate into and out of the probed region current fluctuations result; their autocorrelation function is simply related to that of the adsorbate number fluctuations which depend on D. Comparison of the experimental and theoretical correlation curves yields both D and mean-square fluctuations. A number of systems have been examined in this way and led to the discovery of tunneling, i.e., temperature independent diffusion for H and its isotopes.[7]

Both field emission and field ionization in liquids have been observed and serve to explain breakdown in liquid dielectrics. Under space-charge conditions, electron and ion mobilities in liquids can also be determined from such measurements.[8]

Finally some technological applications. Field emitters are now widely used as electron sources in scanning electron microscopy and electron beam lithography, because they are virtual (almost) point sources free from space-charge limitations. High-density two-dimensional arrays of emitters can be fabricated by various ingenious techniques and are finding a variety of applications.[9] Last, but not least, scanning tunneling microscopy is based on field emission.

Field ion microscopy has been revolutionized by the advent of channel plates which have made it possible to obtain bright images even with heavy imaging gases. This has made it practical to study the surface diffusion of single metal atoms and clusters in great detail.[10] This is still a very active area of surface research. Metallurgical, and for that matter, diffusion studies have been greatly aided by the invention of the atom probe by E. W. Müller.[11] This consists of a field ion microscope with a small hole in the channel plate and screen, and has a means of moving the emitter so as to place a desired region corresponding to as little as a single atom over this hole, which is followed by a drift tube and a detector. A field just insufficient to cause field desorption is applied, followed by a very short additional pulse leading to desorption. The time between the pulse and the arrival of the ion at the detector is then determined and gives the ion's mass to charge ratio. The basic idea has been vastly elaborated and used in studies of alloy surface composition and even in diffusion studies.

Some technological applications of field ionization and desorption should also be mentioned. Single and multiple emitter ion sources for mass spectrometry have been developed and used,[12] most recently in the form of two-dimensional arrays of hollow microcones or craters.[13] Field desorption also forms the basis of liquid-metal ion sources[14] in which electrostatic forces pull a liquid-metal film at the end of a quite blunt solid substrate needle into a Taylor cone with a very fine jetlike protrusion at its end from which field desorption occurs. These sources have high brightness and are also virtual quasi-point sources and are finding important applications in semiconductor technology.

Rereading this book brought back some of the excitement, anguish, and pleasure of the work on which it was in part based, and of the satisfaction of writing it as well as memories of many friends and colleagues. I hope the reader of this reissue will enjoy it.

References

In compiling these references an attempt to concentrate on review articles and monographs has been made. These have an asterisk following the entry.

[1]A. Modinos, *Field, Thermionic, and Secondary Electron Emission Spectroscopy* (Plenum, New York, 1984).*

[2]R. Gomer, *Chemisorption on Metals, Solid State Physics, 30,* 93 (1975).*

[3]J. Kreutzer, in *Chemistry and Physics of Solid Surfaces VIII,* R. Vanselow and R. Howe, eds. (Springer, Berlin, 1990), p. 133.*

[4]M. J. Southon, Ph.D. thesis University of Cambridge, 1963; also in *Field Ion Microscopy,* J. J. Hren and S. Ranganathan, eds. (Plenum, New York, 1968), p. 28.*

[5]C. Wang and R. Gomer, *Surf. Sci. 91,* 533 (1980).

[6]M. Isaacson and R. Gomer, *Appl. Phys. 15,* 253 (1978).

[7]R. Gomer, *Reports on Progress in Physics 53,* 917 (1990).*

[8]R. Gomer, *Accounts of Chem. Research 5,* 41 (1972).*

[9]C. Spindt, C. E. Holland, A. Rosengreen, and I. Brodie, *IEEE Transactions on Electron Devices ED-38,* 2355 (1991).

[10]D. W. Bassett, in *Surface Mobilities on Solid Materials,* V. T. Binh, eds. (Plenum, New York, 1983), p. 63; T. T. Tsong, *ibid.,* p. 109.*

[11]E. W. Müller and T. T. Tsong, *Field Ion Microscopy, Principles and Applications* (Elsevier, New York, 1969).*

[12]H. D. Beckey, *Field Ionization Mass Spectrometry* (Pergamon, Oxford, 1971)*; H. D. Beckey, *Principles of Field Ionization and Field Desorption Mass Spectrometry* (Pergamon, Oxford, 1977).*

[13]C. Spindt, *Surf. Sci. 266,* 145 (1992).

[14]L. W. Swanson and A. Bell, in *The Physics and Technology of Ion Sources,* I. G. Brown, ed. (Wiley, New York, 1989), p. 313.

Author's Preface to the Original Edition

This short book is the outgrowth of four lectures presented by the author in the Department of Chemistry at Harvard University in March 1958. It attempts to present the theory of field emission, field ionization, and field desorption in simple form and to serve as an introduction to field and ion microscopy. The discussion is held at an elementary mathematical level, with emphasis on physical significance rather than rigor. Quantum mechanics is used so sparingly that the main arguments can be understood without previous knowledge of that subject.

A chapter on selected applications, taken largely from the author's research, has been included to illustrate the potentialities of the subject. A description of experimental techniques is presented in appendix form and should be an adequate guide for newcomers to this field.

The number of people responsible for the genesis of any piece of research or writing is probably much larger than they or the author realize. I shall therefore content myself with thanking my colleagues and friends, at the University of Chicago and elsewhere, whose stimuli and interactions have made this book possible.

R. G.

CONTENTS

CHAPTER 1

Theory of Field Emission

FIELD EMISSION AND FIELD IONIZATION

Field emission [1] is defined as the emission of electrons from the surface of a condensed phase into another phase, usually a vacuum, under the action of high (0.3–0.6 v/A) electrostatic fields. This book deals chiefly with emission from metal–vacuum interfaces. The phenomenon consists of the tunneling of electrons *through* the deformed potential barrier at the surface. Thus it differs fundamentally from thermionic emission or photoemission, where only electrons with sufficient energy to go *over* the potential barrier are ejected. However, as in the latter cases, the details of the surface potential configuration are important, so that surface conditions affect field emission profoundly. For this reason one of its chief applications is the study of a variety of surface phenomena.

The closely related phenomenon of field ionization consists of electron tunneling from atoms or molecules under the action of even higher electric fields (2–5 v/A) than required for field emission, and is treated in detail in Chapter 3.

ELECTRONIC PROPERTIES OF SOLIDS

The preceding paragraph indicates that field emission is a quantum-mechanical phenomenon. [2] Fortunately, its theory can be described in fairly elementary terms without compromising our aims. Thus it will be adequate to describe metals by the free-electron model and semiconductors by elementary band theory. [3] A very brief sketch of these follows.

Electron Bands

In a solid the potential wells corresponding to individual atoms or molecules overlap considerably. Consequently outer electrons originally belonging to one atom can tunnel from well to well, and become effectively nonlocalized within the solid. More detailed considerations show that the atomic energy levels coalesce into bands, separated by forbidden regions corresponding to the gaps between atomic levels. Each band has a fine structure, corresponding to and equal in number to the original atomic levels.

Electrons are fundamental particles of spin $\frac{1}{2}$ and obey the Pauli exclusion principle, so that each translational level can accommodate no more than two electrons (spin $\frac{1}{2}$ and $-\frac{1}{2}$). If it happens that all bands are completely filled by electrons, the solid will be an insulator, since there will be no allowed empty levels into which electrons may be raised by an applied field. If there are one or more partially filled bands, the solid will be a metal, since electrons can be accelerated by small fields and transport charge.

Sometimes a completely empty band dips close to a filled one, so that the gap separating them is small. Under these conditions electrons from the full band may be thermally excited into the empty one, so that conduction can occur. Such a substance is known as an intrinsic semiconductor. It is often possible to include impurity atoms of different valence in a semiconductor. These atoms "feel" the dielectric constant of their host (unless they are very small) and consequently have very much reduced ionization potentials for electrons or positive holes. Thus they will either contribute electrons to an empty band or accept them from a full one. Either case leads to conduction. Semiconductors containing electron donors are known as n-type (negative), while those containing acceptors are called p-type (positive.)

For many purposes it is adequate to consider only one partially filled (conduction) band, and to treat this as if it corresponded to electrons contained in a rectangular three-dimensional potential well. This ignores both the corrugation and the periodicity of the potential structure giving rise to the band. The treatment will, therefore, be least adequate for electrons of low energy which feel these effects most strongly. However, electrons near the top of the band will behave in many respects like free electrons.

Fermi Statistics

Each translational level or quantum cell can accommodate only two electrons, so that these obey Fermi statistics.[3] Their behavior differs in many important respects from that of Boltzmann particles, which they resemble only at much greater dilution than is encountered in the conduction bands of metals.

When electrons are stacked in the available levels until all of them have been disposed of, the highest filled level (known as the Fermi level μ) is several electron volts above the bottom of the band in all metals. Since this energy corresponds to a temperature of $\sim 10^{4\circ}$ K, it is obvious that there will be very little excitation above the Fermi level at ordinary temperatures. Conversely, there can be no condensation into the ground state even at $0°$ K. Furthermore, there are many more quantum cells near the Fermi level than near the bottom of the band, so that the average energy of the electron gas will be close to μ, even at $0°$ K. It is this fact which makes field emission virtually temperature independent.

Since there is negligible excitation above μ at low temperatures, it constitutes an effective ceiling on electron energies. Therefore, the distribution of momenta among the Cartesian degrees of freedom cannot be independent, as in a Boltzmann gas, but is subject to the restriction that the total energy must not exceed μ (at $0°$ K). This turns out to be important in calculating the resolution of a field-emission microscope.

Work Function

The solid-state analogue of the ionization potential in atoms or molecules is called the work function, ϕ, and corresponds to the energy difference between μ and a field-free vacuum near the surface (Fig. 1a).

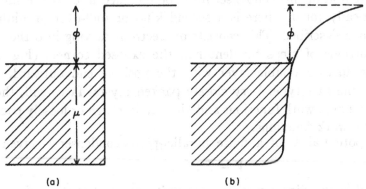

Fig. 1. Potential-energy diagram for electrons in a metal: μ, Fermi energy; ϕ, work function; (a) abrupt potential step at the surface; (b) image potential included.

The work function ϕ has values of 2–5 v for metals and arises from two effects. The first, or inner potential, is the intrinsic partial free energy of "solution" of electrons in the metal and is the difference between the chemical potential of the electron in the metal and that at a very large distance from it in a field-free vacuum. The second component arises from electrostatic effects at the surface and decays slowly with distance. This can vary with surface structure and is responsible for the crystallographic anisotropy in ϕ, which exists even on perfectly clean surfaces.[4]

Attempts to calculate the inner potential from first principles [5] are in a rather rudimentary state, since the problem is even more complicated than the closely related one of calculating binding energies. Even less quantitative theoretical work has been done on the electrostatic component, although a qualitatively correct and physically very illuminating treatment has been given by Smoluchowski [6] to account at least for the anisotropy in this term.

Double Layers

If a metal surface is completely planar, the electron cloud in the interior will not terminate there abruptly, since this would correspond to an infinite gradient of the wave function and hence to infinite kinetic energy. Instead, there will be a gradual decay, as shown in Fig. 2a, with a Debye length of 0.5–1 A. This spill-over causes an electron deficiency on the metal side of the interface and gives rise to a double layer or condenser, with the negative end outermost. The potential of this condenser must be included in ϕ. If the surface is atomically rough, there is a second kind of spill-over in addition to the one described. This consists of electrons flowing into the concave portions of steps by denuding the exposed corners (Fig. 2b). This produces a double layer with the positive end outward, and opposes the first type. Very closely packed crystal faces will, therefore, have high work functions and atomically rough or loosely packed ones, low work functions.

The potential V of a finite parallel-plate condenser is given by

$$V = P\Omega \leqslant 4\pi P, \qquad (1)$$

where P is the dipole moment per unit area and Ω the solid angle subtended at the point of observation. At the condenser plate Ω is

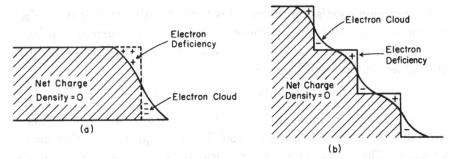

Fig. 2. Charge distribution at a metal surface (schematic): (*a*) atomically smooth surface; (*b*) atomically stepped surface.

4π, and it decays as $1/r^2$, so that all electrostatic contributions to ϕ vanish at large distances for finite crystals. All the phenomena of field emission occur so near the surface that the decay can be ignored, apart from reassuring us that the first law of thermodynamics has not been violated.

Surface double layers arise also from adsorption. Changes in the charge distribution occur in such a way that a dipole moment P_i (in first approximation) can be associated with each adsorbed particle (henceforth: ad-particle). The adsorbed layer will, therefore, contribute a term $\Delta\phi$ to the work function:

$$\Delta\phi = 2\pi P_i N_s \theta, \tag{2}$$

where N_s is the maximum number of adsorption sites per unit area and θ the fraction of filled ones. Equation (2) is correct as written if the dipole is centered symmetrically about the (imaginary) surface of electroneutrality, since an emerging electron must then do work against half the layer only. If the dipole is completely contained in the ad-particle, the right-hand member of Eq. (2) must be multiplied by 2. In general, it is more plausible to assume the former. In practice, the measured quantity is $\Delta\phi$. This is often 1–2 volts, even for physically adsorbed gases on clean metals.

Thermionic Emission and Photoemission

We are now in a position to examine the emission of electrons from metals. Although our main interest is the high-field case, a few words on thermionic emission and photoemission are appropriate by way of contrast.

In thermionic emission the metal is heated until a sufficient number of electrons acquire kinetic energies $\geqslant \phi + \mu$. The current density i can be shown (most simply by assuming the rate of emission to equal that of arrival at equilibrium [7]) to be

$$i = 120T^2 e^{-\phi/kT} \text{ amp/cm}^2. \tag{3}$$

The work function of most high-melting metals is 4–5 volts. Therefore temperatures upward of 1500° K are required for thermionic emission. This constitutes its chief drawback as a tool for the study of adsorption, since few if any systems can be heated to such high temperatures without radical changes. Nevertheless, thermionic emission has proved very useful as a method of obtaining the work functions of clean high-melting metals and has practical applications too well known to require mention.

In photoemission,[8] surfaces are irradiated with light of energy $h\nu \geqslant \phi$. The threshold wavelength for clean metals lies in the visible or near ultraviolet; very little temperature dependence is to be expected. Photoemission is thus a powerful tool for surface studies.

FIELD EMISSION FROM METALS

Tunneling of Electrons

In thermionic emission and photoemission electrons are given sufficient energy to overcome the potential barrier at the metal surface. In field emission, on the other hand, the barrier is deformed so strongly that unexcited electrons can leak out *through* it. The situation is illustrated in Fig. 3a. When a field F is applied to the metal surface, electrons of kinetic energy E_x along the emission direction see a barrier of height $\phi + \mu - E_x$ and thickness $(\phi + \mu - E_x)/Fe$. If this is thin and low enough, penetration will occur with finite probability.

Tunneling is a purely quantum-mechanical phenomenon with no classical analogue. It can best be reconciled with macroscopic intuition by the Heisenberg uncertainty principle. A knowledge of the momentum of an electron within an uncertainty Δp implies a corresponding uncertainty Δx in its position, given by

$$\Delta p \cdot \Delta x \cong \hbar/2, \tag{4}$$

where $\hbar = h/2\pi$, h being Planck's constant. If we consider electrons

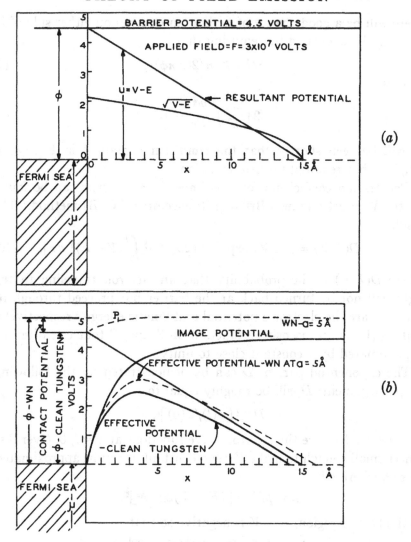

Fig. 3. Potential-energy diagram for electrons at a metal surface in the presence of an applied field: (a) without image potential; (b) with image potential included; broken curves refer to the additional potential caused by adsorption of nitrogen.

near the Fermi level, the pertinent uncertainty in momentum is that corresponding to the barrier of height ϕ, $(2m\phi)^{\frac{1}{2}}$. If the corresponding uncertainty in position,

$$\Delta x \cong \hbar/2(2m\phi)^{\frac{1}{2}}, \qquad (5)$$

is of the order of the barrier width,

$$x = \phi/Fe, \qquad (6)$$

there will be a good chance of finding an electron on either side of it. This can be expressed by requiring that

$$\phi/Fe \cong \hbar/2(2m\phi)^{\frac{1}{2}}, \tag{7}$$

or that

$$2\left(\frac{2m}{\hbar^2}\right)^{\frac{1}{2}} \frac{\phi^{\frac{3}{2}}}{Fe} \approx 1. \tag{8}$$

It will be seen shortly that the condition contained in Eq. (8) is roughly that required for field emission.

Penetration coefficients for one-dimensional barriers can be found by the Wentzel-Kramers-Brillouin (henceforth: WKB) method.[9] The result is

$$D(E,V) = f(E,V) \exp\left[-2(2m/\hbar^2)^{\frac{1}{2}} \int_0^l (V-E)^{\frac{1}{2}}\, dx\right], \tag{9}$$

where $D(E,V)$ is the probability that an electron traveling to the right will not be turned back at the barrier but proceed through it, V and E are the electron's potential and kinetic energies respectively, and $f(E,V)$ is an insensitive function of V and E that can often be approximated by a constant close to unity.

The exponential part of D can be demonstrated by the following simple argument: D will be roughly equal to

$$D \cong [\psi(l)/\psi(0)]^2,$$

where $\psi(l)$, $\psi(0)$ are the electron wave functions at $x = l$, 0. Over distances small enough to make V seem constant, ψ can be approximated by e^{ikx}, where k is

$$k = p/\hbar = [(E-V)2m/\hbar^2]^{\frac{1}{2}}.$$

In the barrier region, $E - V$ is negative so that

$$\psi = \exp\left[-(2m/\hbar^2)^{\frac{1}{2}}(V-E)^{\frac{1}{2}}x\right].$$

Therefore

$$\psi(x+dx)/\psi(x) = \exp\left[-(2m/\hbar^2)^{\frac{1}{2}}(V-E)^{\frac{1}{2}}dx\right]$$

and

$$\psi(l)/\psi(0) = \exp\left[-(2m/\hbar^2)^{\frac{1}{2}} \int_0^l (V-E)^{\frac{1}{2}}\, dx\right],$$

so that

$$D \cong \exp\left[-2(2m/\hbar^2)^{\frac{1}{2}} \int_0^l (V-E)^{\frac{1}{2}}\, dx\right].$$

Examination of Fig. 3a shows that the exponent in the right-hand member of Eq. (9) has a simple meaning. It represents, apart from the factor $2(2m/\hbar^2)^{\frac{1}{2}}$, the area under the curve with ordinate $(V - E)^{\frac{1}{2}}$ from $x = 0$ to $x = l$. This area is nearly triangular in shape, with base $(\phi + \mu - E_z)/Fe$ and altitude $(\phi + \mu - E_z)^{\frac{1}{2}}$, so that it is given by

$$A \cong \tfrac{1}{2}(\phi + \mu - E_z)^{\frac{3}{2}}/Fe. \tag{10}$$

Substitution in Eq. (9) yields for D

$$D = f(E_z, V) \exp\left[-(2m/\hbar^2)^{\frac{1}{2}}(\phi + \mu - E_z)^{\frac{3}{2}}/Fe\right]. \tag{11}$$

This is, in fact, very similar to D calculated more rigorously by Fowler and Nordheim:[2]

$$D = \frac{4[E_z(\phi + \mu - E_z)]^{\frac{1}{2}}}{(\phi + \mu)} \exp\left[-\frac{4}{3}\left(\frac{2m}{\hbar^2}\right)^{\frac{1}{2}}\frac{(\phi + \mu - E_z)^{\frac{3}{2}}}{Fe}\right]. \tag{12}$$

If we limit ourselves to electrons of $E_z \cong \mu$, Eq. (12) yields

$$D = \frac{4(\phi\mu)^{\frac{1}{2}}}{(\phi + \mu)} \exp\left[-6.8 \times 10^7 \frac{\phi^{\frac{3}{2}}}{F}\right], \tag{13}$$

for ϕ in volts and F in volts per centimeter. Since most electrons have energies near μ, the emitted current can be approximated by multiplying Eq. (13) by the total rate of arrival of electrons. A more accurate procedure consists in using Eq. (12) for the penetration probability, multiplying it by the appropriate differential arrival rate, and integrating over $0 \leqslant E_z \leqslant \mu$, as is done on page 19. The result is the Fowler-Nordheim equation:[2]

$$i = 6.2 \times 10^6 \frac{(\mu/\phi)^{\frac{1}{2}}}{\mu + \phi} \cdot F^2 \exp\left[-6.8 \times 10^7 \frac{\phi^{\frac{3}{2}}}{F}\right] \text{ amp/cm}^2, \tag{14}$$

for all energies in electron volts and F in volts per centimeter.

Equation (14) shows that current densities of 10^2–10^3 amp/cm^2 can be expected for fields of 3–6×10^7 v/cm. This is the range most frequently occurring in emission work.

Image Effect

Equations (10–14) are correct for an abrupt potential step at the metal surface. We have already seen that the charge cloud does not terminate sharply so that a vertical rise is not to be expected. Besides, the potential near the surface is decreased by an image term V_{im}

$$V_{im} = -e^2/4x = 3.6/x,\qquad(15)$$

for V in volts and x in angstroms. The classical expression (15) undoubtedly breaks down near the surface and must be replaced by appropriate exchange- and correlation-energy terms. Since no accurate calculations of the electron potential at a surface exist,[5] the use of the classical image correction is the best one can do. In the presence of an applied field F, the potential at a metal surface has the form

$$V = -Fex - e^2/4x\qquad(16)$$

if the zero point is taken at $\phi + \mu$ above the bottom of the conduction band. Figure 3b shows that the resultant barrier is smaller than that obtained by neglecting image effects. The position and value of the top of this barrier (Schottky saddle [10]) can be found by setting the derivative of Eq. (16) equal to zero. The result is

$$x_{max} = (3.6/F)^{\frac{1}{2}}\qquad(17)$$

and

$$V_{max} = -3.8F^{\frac{1}{2}}\qquad(18)$$

for x in angstroms and F in volts per angstrom.

It is now a simple matter to estimate the effect of the image term on field emission. The requisite area is again almost triangular in shape, with base ϕ/F and altitude $(\phi - 3.8 \times 10^{-4}F^{\frac{1}{2}})^{\frac{1}{2}}$ if F is measured in volts per centimeter. Its area is

$$A = \tfrac{1}{2}(\phi/F)(\phi - 3.8 \times 10^{-4}F^{\frac{1}{2}})^{\frac{1}{2}}$$
$$= \tfrac{1}{2}(\phi^{\frac{3}{2}}/F)(1 - 3.8 \times 10^{-4}F^{\frac{1}{2}}/\phi)^{\frac{1}{2}}.\qquad(19)$$

It is seen that this differs from the uncorrected area by the multiplicative factor

$$\alpha = (1 - y)^{\frac{1}{2}},\qquad(20)$$

where

$$y = 3.8 \times 10^{-4}F^{\frac{1}{2}}/\phi.\qquad(21)$$

A more accurate calculation by Nordheim [11] results in a correction term differing from our result only by an almost constant factor of 1.5.

The image correction has also been treated by Burgess et al.[1] who evaluated the WKB integral analytically, using the correct potential of Eq. (16). As the preceding discussion would suggest, all of these results take the form of multiplicative corrections to the Fowler-Nordheim exponent, $\alpha(y)$, which are functions of the argument y defined by Eq. (21). The results of the various calculations are summarized in Table 1.

TABLE 1. Image correction for exponent of Fowler-Nordheim equation.

y	0	0.2	0.3	0.4	0.5	0.6	0.7	0.8	0.9	1.0
α_N	1	0.951	0.904	0.849	0.781	0.696	0.603	0.494	0.395	0
$(1 - y)^{\frac{1}{2}}$	1	0.894	0.836	0.774	0.707	0.632	0.548	0.447	0.316	0
α_B	1	0.937	0.872	0.789	0.690	0.577	0.450	0.312	0.161	0

$y = 3.8 \times 10^{-4} F^{\frac{1}{2}}/\phi$; α_N = Nordheim correction;[11] α_B = Burgess correction [1]

Under most of the conditions encountered in practice, y is 0.2–0.35, so that reductions in applied field of the order of 10–20 percent result from the image effect. Since α is a slowly varying function of F, it is generally permissible to treat it as a constant in a given application of the Fowler-Nordheim equation.

Graphic Evaluation of the Penetration Coefficient

In addition to the analytic procedures outlined above, it is often useful to evaluate special barriers graphically. This can always be done by recalling the geometric meaning of the exponent in Eq. (9).

ENERGY DISTRIBUTION OF FIELD EMISSION

The energy distribution of emitted electrons is important in the derivation of the Fowler-Nordheim equation, the resolution of the field-emission microscope, and the noise characteristics of field-emission cathodes. It is also of considerable intrinsic interest, as will be seen.

The number of electrons in unit volume in the momentum range $dp_x \, dp_y \, dp_z$, namely $N(p_x,p_y,p_z) \, dp_x \, dp_y \, dp_z$, is given by the number of cells in the corresponding volume of phase space multiplied by the Fermi-Dirac distribution function:

$$N(p_x, p_y, p_z) \, dp_x \, dp_y \, dp_z = (2/h^3)(1 + e^{(E-\mu)/kT})^{-1} \, dp_x \, dp_y \, dp_z. \quad (22)$$

The factor 2 in the right-hand member arises from the electron spin. At $0°$ K the Fermi function $(1 + e^{(E-\mu)/kT})^{-1}$ takes the value 1 for $E \leqslant \mu$ and 0 otherwise. This is merely another way of saying that all levels up to but not beyond μ are filled at $0°$ K. In terms of velocities, v_x, v_y, v_z, Eq. (22) takes the form

$$N(v_x, v_y, v_z)\, dv_x\, dv_y\, dv_z = 2(m/h)^3(1 + e^{(E-\mu)/kT})^{-1}\, dv_x\, dv_y\, dv_z, \tag{23}$$

with

$$E = \tfrac{1}{2}m(v_x^2 + v_y^2 + v_z^2). \tag{24}$$

Distribution Normal to the Emission Surface

We calculate first the number of electrons with velocities along the emission direction x in the range between v_x and $v_x + dv_x$ regardless of the values of v_y and v_z. This number, $N(v_x)\, dv_x$, is found by integration over dv_y and dv_z after changing to polar coordinates in the v_{yz} plane.

$$N(v_x)\, dv_x = \frac{4\pi m^2}{h^3}\, dv_x \int_0^\infty (1 + e^{(E_x + \epsilon - \mu)/kT}) - 1\, d\epsilon$$

$$= \frac{4\pi m^2 kT}{h^3}\, \ln(1 + e^{\Delta/kT})\, dv_x, \tag{25}$$

where $\Delta = \mu - E_x$ and $E_x = \tfrac{1}{2}mv_x^2$.

At sufficiently low temperature $|\Delta/kT| \gg 1$, so that

$$kT \ln(1 + e^{\Delta/kT}) \rightarrow \begin{cases} 0 \text{ if } E_x > \mu & (\Delta \text{ negative}); \\ \Delta \text{ if } E_x \leqslant \mu & (\Delta \text{ positive or zero}). \end{cases} \tag{26}$$

The low-temperature form of $N(v_x)\, dv_x$ therefore becomes

$$N(v_x)\, dv_x = \frac{4\pi m^2 \Delta}{h^3}\, dv_x. \tag{27}$$

At moderate temperatures this result is still valid when $E_x < \mu$, but the logarithm no longer vanishes when $E_x > \mu$ and $\Delta < 0$. In that case its expansion yields

$$\ln(1 + e^{\Delta/kT}) \cong e^{\Delta/kT} \qquad \text{for } e^{\Delta/kT} \ll 1, \tag{28}$$

so that, at moderate T

$$N(v_x)\, dv_x \cong \frac{4\pi m^2 kT}{h^3}\, e^{\Delta/kT}\, dv_x \qquad \text{for } E_x > \mu \tag{29a}$$

and

$$N(v_x)\, dv_x \cong \frac{4\pi m^2 \Delta}{h^3}\, dv_x \qquad \text{for } E_x < \mu. \qquad (29b)$$

The energy distribution normal to the surface, $I(E_x)\, dE_x$, can now be found as the product of the flux $v_x N(v_x)\, dv_x$ and the barrier penetration coefficient $D(E_x)$, given by Eq. (12). Since

$$dE_x = mv_x\, dv_x, \qquad (30)$$

the result in terms of energy is:

$$\begin{aligned}
&I(E_x)\, dE_x \\
&= \frac{16\pi m [E_x(\phi + \Delta)]^{\frac{1}{2}}}{h^3(\phi + \mu)} \ln\left(1 + e^{\Delta/kT}\right) \exp\left[-b(\phi + \Delta)^{\frac{3}{2}}/F\right] dE_x, \quad (31)
\end{aligned}$$

where

$$b = 6.8 \times 10^7 \alpha \qquad (32)$$

and α is the image correction term. At moderate temperatures this becomes

$$I(E_x)\, dE_x = \frac{16\pi m \Delta [E_x(\phi + \Delta)]^{\frac{1}{2}}}{h^3(\phi + \mu)} \exp\left[-b(\phi + \Delta)^{\frac{3}{2}}/F\right] dE_x$$

$$\text{for } E_x < \mu \quad (33a)$$

and

$$I(E_x)\, dE_x = \frac{16\pi m k T [E_x(\phi + \Delta)]^{\frac{1}{2}} e^{\Delta/kT}}{h^3(\phi + \mu)} \exp\left[-b(\phi + \Delta)^{\frac{3}{2}}/F\right] dE_x$$

$$\text{for } E_x > \mu. \quad (33b)$$

Equation (33a) also holds for the pure field-emission or $0°$ K case. Equations (33a) and (33b) show clearly that $I(E_x)\, dE_x$ depends only very slightly on the absolute value of μ but strongly on $\Delta = \mu - E_x$ and on ϕ, since these two quantities determine the barrier dimensions.

Figure 4 shows graphs of $I(E_x)$ as a function of E_x for three values of applied field at $300°$ K. It is seen that the distribution broadens considerably as the field is increased, and that its maximum shifts to the left. This is due to the fact that increases in field thin the barrier and permit electrons of lower E_x to escape. The curves also show the contribution of electrons with $E > \mu$ to the total current.

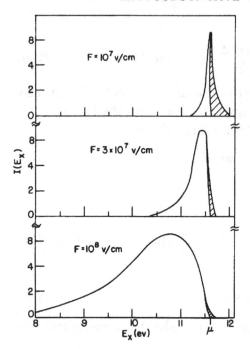

Fig. 4. Theoretical normal energy distribution of field-emitted electrons. The curves have been drawn to make the ordinates comparable. The shaded areas refer to $E_x > \mu$ at 300° K.

At 300° K this "Boltzmann tail" is important only at very low fields which place a premium on electron energy, regardless of low abundance.

Total Energy Distribution

Attempts to measure the normal energy distribution in field emission experimentally have so far met with failure. Relatively high voltages are required for emission, so that even very small imperfections in the geometry of the experimental system will cause a sufficient transfer of energy from the normal to the tangential directions to smear out the theoretical distribution. However, it is possible to devise experimental geometries in which the total energy distribution is measured. The latter also determines the noise spectrum of a field emitter. It is therefore interesting to calculate it, as was recently done by R. Young.[12]

We start by determining the number of electrons with total energy $E = \frac{1}{2}mv^2$, striking a surface element dS in unit time. We define θ as the angle between the surface normal and the total velocity vector v, and φ as the azimuthal angle measured in the surface plane. The volume contributing electrons of velocity v from a solid-angle element $d\Omega$ at θ and φ is $dS\, v \cos \theta$. The number of electrons with corresponding energies between E and $E + dE$ contained in this volume is $N(E)\, dE\, dS\, v \cos \theta$, where $N(E)\, dE$ is the number of electrons per unit volume in this energy range. Of this number only the fraction corresponding to electrons with v vectors directed at dS will actually impinge there. This fraction is given by $(1/4\pi)$ times the element of solid angle subtended in velocity space by the vectors of magnitude $|\,v\,|$ pointing at dS. This element is equal to $d\Omega$, the element of solid angle in real space at θ and φ, and is given by $\sin \theta\, d\theta\, d\varphi$. Consequently the flux arriving at dS from $d\Omega$ is

$$J(E,\Omega)\, dE\, d\Omega = \frac{1}{4\pi} N(E)\, dE\, dS\, v \cos \theta \sin \theta\, d\theta\, d\varphi. \qquad (34)$$

Integration over φ can be carried out at once, yielding for the differential flux per unit area of surface

$$J(E,\theta)\, dE\, d\theta = \tfrac{1}{2}N(E)\, v \cos \theta \sin \theta\, d\theta\, dE. \qquad (35)$$

Since θ is a measure of E_x, the surface normal energy component, for a fixed E, Eq. (35) can be transformed into an equivalent one in terms of E and E_x. For fixed E,

$$dE_x = d(\tfrac{1}{2}mv^2 \cos^2 \theta) = -mv^2 \cos \theta \sin \theta\, d\theta, \qquad (36)$$

so that

$$J(E,E_x)\, dE\, dE_x = -\tfrac{1}{2}(2mE)^{-\frac{1}{2}} N(E)\, dE\, dE_x. \qquad (37)$$

Here $N(E)\, dE$ is the usual Fermi-Dirac function in terms of energy,

$$N(E)\, dE = \frac{4\pi}{h^3} \frac{(2m)^{\frac{3}{2}}E^{\frac{1}{2}}}{1 + e^{(E-\mu)/kT}}\, dE, \qquad (38)$$

so that

$$J(E,E_x)\, dE\, dE_x = \frac{4\pi m}{h^3} (1 + e^{(E-\mu)/kT})^{-1}\, dE\, dE_x. \qquad (39)$$

The total energy distribution of emitted electrons can now be found by multiplying this result by the barrier penetration coefficient, $D(E_x)$, and integrating over the normal energy range dE_x:

$$I(E)\, dE = - \frac{16\pi m\, dE}{h^3(\phi + \mu)(1 + e^{(E-\mu/kT)})}$$

$$\times \int_{E_x = E}^{E_x = 0} [E_x(\phi + \Delta)]^{\frac{1}{2}} \exp [-b(\phi + \Delta)^{\frac{3}{2}}/F]\, dE_x. \qquad (40)$$

The limits of integration as written correspond to those of θ from 0 to $\pi/2$.

The integral in Eq. (40) can be simplified in view of the following considerations. At ordinary temperatures, emission comes mainly from the vicinity of the Fermi level, that is, $E_x \cong \mu$, so that Δ is small compared to, say, ϕ. Under these conditions the factor $[E_x(\phi + \Delta)]^{\frac{1}{2}}$ is nearly constant and takes the value

$$[E_x(\phi + \Delta)]^{\frac{1}{2}} \approx (\phi\mu)^{\frac{1}{2}}. \tag{41}$$

The image correction term contained in b, Eq. (32), can also be considered independent of E_x and given by Eq. (20). Further, we may expand the Fowler-Nordheim exponent as follows:

$$(b/F)(\phi + \Delta)^{\frac{3}{2}} = (b/F)\phi^{\frac{3}{2}}(1 + \Delta/\phi)^{\frac{3}{2}} \cong (b/F)\phi^{\frac{3}{2}} + (b/F)\tfrac{3}{2}\phi^{\frac{1}{2}}\Delta, \tag{42}$$

since

$$\Delta/\phi \ll 1. \tag{43}$$

This reduces the integral to the form $\int e^{cy}dy$. Since the region $E_x = 0$ contributes effectively nothing to the emission, we may replace this limit by $-\infty$, obtaining

$$I(E) \, dE = \frac{16\pi m(\phi\mu)^{\frac{1}{2}}e^{-b\phi^{\frac{3}{2}}/F}}{h^3(\phi + \mu)(1 + e^{(E-\mu)/kT})} \, dE$$

$$\times \int_{-\infty}^{E} \exp\left[\tfrac{3}{2}b\phi^{\frac{1}{2}}(E_x - \mu)/F\right] dE_x. \tag{44}$$

Integration finally yields

$$I(E) \, dE = \frac{32\pi m\mu^{\frac{1}{2}}F \exp\left(-b\phi^{\frac{3}{2}}/F\right)}{3h^3b(\phi + \mu)} \cdot \frac{\exp\left[(\tfrac{3}{2}\phi^{\frac{1}{2}}b/F)(E - \mu)\right]}{1 + \exp\left[(E - \mu)/kT\right]} \, dE. \tag{45}$$

Equation (45) can be brought into explicit agreement with Young's formula by choosing field-free vacuum instead of the bottom of the conduction band as the zero of energy. From this reference point the total electron energy becomes the negative quantity

$$E' = E - (\phi + \mu), \tag{46}$$

so that

$$E - \mu = E' + \phi \tag{47}$$

and

$$I(E') \, dE' = \frac{32\pi m\mu^{\frac{1}{2}}F \exp\left(\tfrac{1}{2}b\phi^{\frac{3}{2}}/F\right)}{3h^3b(\phi + \mu)} \cdot \frac{\exp\left(\tfrac{3}{2}b\phi^{\frac{1}{2}}E'/F\right)}{1 + \exp\left[(E' + \phi)/kT\right]} \, dE'. \tag{48}$$

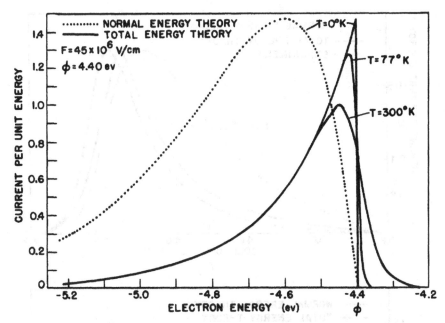

Fig. 5. Total and normal energy distributions for field-emitted electrons.
After Young.[12]

Figure 5 shows calculated values of $I(E')$ at various temperatures. It is seen that this distribution is considerably narrower than the corresponding normal one, which is also shown, and that its maximum is fairly temperature-sensitive. Inspection of Eq. (48) shows that the total-energy distribution has the form of a Fermi-Dirac function, modulated by barrier-penetration coefficients. The leading edge of the Fermi distribution (where E' is positive, or at any rate small) is rather accurately reproduced by $I(E')$. This accounts for the temperature dependence of its maximum, since the "Boltzmann tail" appears as a corresponding deficiency in the distribution on the other side of μ.

It can be shown that the half width σ of the total-energy distribution at 0° K is

$$\sigma = 0.46F/b. \qquad (49)$$

Young and Müller [13] have recently succeeded in measuring $I(E)$ experimentally as a function of temperature. As Fig. 6 shows, the agreement with the calculated curves is remarkably good. This indicates that the use of the free-electron model in the theory of field emission is quite justified.

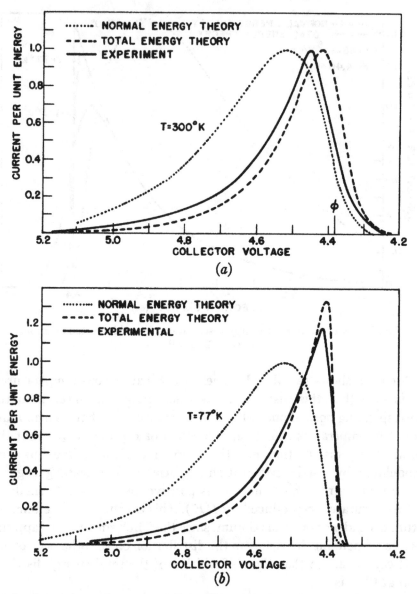

Fig. 6. Comparison of experimental and theoretical total energy distributions in field emission, taken from Young and Müller: [13] (a) $T = 300°$ K; (b) $T = 77°$ K.

THE FIELD-EMISSION EQUATION

The expression for total current density i can be obtained by integrating the normal-energy distribution over E_x.

0° K *Case*

The "pure" Fowler-Nordheim equation can easily be obtained from Eq. (33a). The approximations used in evaluating the resultant integral are identical to those discussed in connection with Eqs. (40–45) and lead to

$$i = \frac{16\pi m e(\phi\mu)^{\frac{1}{2}}}{h^3(\phi+\mu)} \exp\left(-b\phi^{\frac{3}{2}}/F\right)\int_{-\infty}^{\mu} \exp\left(\tfrac{3}{2}b\phi^{\frac{1}{2}}\Delta/F\right)\Delta\,dE_x. \quad (50)$$

This is of the form $-\int y e^{cv}\,dy$ and results finally in

$$i = \frac{4}{3}\frac{16\pi m e(\mu/\phi)^{\frac{1}{2}}}{h^3(\phi+\mu)b^2} F^2 \exp\left(-b\phi^{\frac{3}{2}}/F\right)$$

$$= 6.2 \times 10^6 \frac{(\mu/\phi)^{\frac{1}{2}}}{\alpha^2(\phi+\mu)} F^2 \exp\left(-6.8 \times 10^7\phi^{\frac{3}{2}}\alpha/F\right). \quad (51)$$

This is the Fowler-Nordheim equation with a first-order image correction.

Moderate Temperatures

At moderate temperatures, where the emission still occurs mainly from the vicinity of the Fermi level (on either side of it), the Fowler-Nordheim penetration coefficient may be expanded as in Eq. (42). Equation (31) can then be integrated analytically,[1] yielding

$$i(T) = i(0)\,\frac{\tfrac{3}{2}\pi k T b\phi^{\frac{1}{2}}/F}{\sin\left[\tfrac{3}{2}\pi k T b\phi^{\frac{1}{2}}/F\right]}. \quad (52)$$

The low-temperature expansion of Eq. (52) gives

$$i(T) = i(0)\left[1 + \tfrac{2}{27}(\pi k T b\phi^{\frac{1}{2}}/F)^2 + \cdots\right]. \quad (53)$$

Equation (52) shows that $i(300° \text{ K}) = 1.03\, i(0° \text{ K}$, while $i(1000° \text{ K}) = 1.5\, i(0° \text{ K})$. Thus the error involved in the use of the "pure" field-emission equation, Eq. (51) or Eq. (14), is very small at ordinary temperatures.

Thermionic-Field Emission

At higher temperatures, where the expansion of D is invalid, the energy distribution can be integrated numerically or graphically. Figure 7 shows the results of such a calculation by Dyke and Dolan,[14] who have termed this the T-F region. Under most conditions, space-

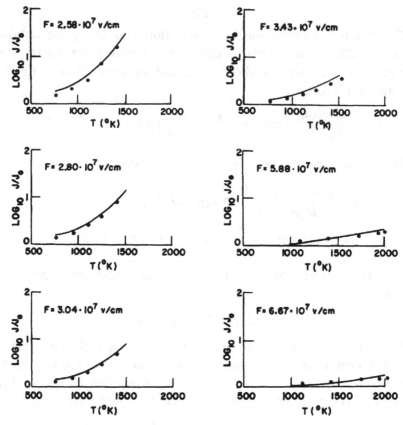

Fig. 7. Calculated (solid curves) and experimental (dots) values of log $[J(T)/J(0)]$ for several values of applied field, from Dyke and Dolan;[14] $J(300)$ and $J(T)$ are current densities at 300° K and T° K respectively.

charge effects tend to set in here and further complicate calculations. The results of Fig. 7 ignore this. It can be seen that increases in field tend to reduce the difference between pure and T-F emission. This results from the fact that, at sufficiently high fields, it does not matter very much whether an electron is above or below the Fermi level.

FIELD EMISSION FROM SEMICONDUCTORS

Although little experimental work has been done on this subject,[15–18] the use of vapor-grown whiskers promises to overcome some of the experimental difficulties connected with cleaning semiconductor surfaces. It is therefore worth while to give a brief account of the theory.

Detailed mathematical analyses have been carried out by Margulis [19] and by Stratton.[20] This section emphasizes the physical situation and presents the theory in simplified form.

Elementary Band Model of Semiconductors

A very brief description of semiconductors will first be presented. It was pointed out in the beginning of this chapter that conduction in a solid hinges on the accessibility of empty levels. In a bona fide metal these levels are always present, since the Fermi level lies below the top but above the bottom of the highest populated (that is, conduction) band. In an insulator, the bands are either completely filled or completely empty. It may happen that an empty band lies only a little above the filled valence band. In that case thermal excitation across the gap places electrons in the almost empty conduction band and creates positive holes in the valence band, so that conduction can occur. A substance of this kind is known as an intrinsic semiconductor (Fig. 8).

In a metal, the Fermi level coincides with the highest filled state at 0° K and changes only very slightly at moderate temperatures. In a semiconductor, the Fermi level lies approximately at the middle of the energy gap. This can be shown as follows. The equilibrium constant for the reaction

$$0 \leftrightarrow e_c + h_v, \tag{54}$$

where e_c is an electron in the conduction band and h_v a hole in the

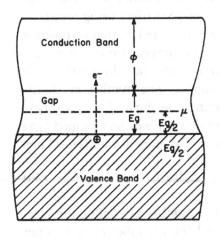

Fig. 8. Schematic diagram of energy bands for an intrinsic semiconductor: E_g, width of gap; μ, Fermi level; ϕ ionization energy from bottom of conduction band.

valence band, can be expressed in terms of the corresponding partition functions Q_i and number densities n_i by

$$K_{equ} = n_e n_h = Q_e Q_h \exp{(-E_g/kT)}, \qquad (55)$$

E_g being the energy gap between the bands. Since the population of conduction electrons and holes is small, Boltzmann statistics are applicable, so that Eq. (55) becomes

$$n_e n_h = 4(2\pi mkT/h^2)^3 \exp{(-E_g/kT)}, \qquad (56)$$

if the hole and electron masses are assumed equal. Consequently,

$$n_e = 2(2\pi mkT/h^2)^{\frac{3}{2}} \exp{(-E_g/2kT)}, \qquad (57)$$

since $n_e = n_h$. It follows from the definition of the partition function, measured from the bottom of the conduction band,

$$Q_e = 2(2\pi mkT/h^2)^{\frac{3}{2}} = n_e e^{-\mu/kT}, \qquad (58)$$

that $\mu = -E_g/2$. It should also be remembered that μ is the chemical potential of the electrons in the system.

The presence of impurity atoms of different valence also leads to semiconductivity. Let us assume that the impurity has an extra valence electron compared with the host. This electron will describe hydrogenlike orbits about the impurity ion core. However, the Coulomb attraction is reduced by the dielectric constant K of the medium, so that the radii are increased by K, while the ionization energy E_i is reduced by K^2. Consequently, thermal ionization into the conduction band will occur very readily. A donor semiconductor of this kind is said to be of n-type, the charge carriers being negative. Acceptor impurity atoms will similarly cause the creation of positive holes in the valence band. The process may be regarded as the ionization of holes from the neutral acceptors, so that the situation is mathematically identical to that for donors. The number of charge carriers can be calculated by considering the donor (or acceptor)–electron (or hole) equilibrium. For simplicity it will be assumed that $E_g \gg E_i$, and that only a small fraction of donors are ionized. Then

$$\frac{n_e n_i}{N} \cong \frac{Q_e Q_i}{Q_N} \exp{(-E_i/kT)}, \qquad (59)$$

where n_e, n_i, and N are respectively the number of conduction electrons, the number of ionized donors, and total number of donors per unit volume. Since Q_i and Q_N differ only in their multiplicities (to a first approximation),

$$n_e = N^{\frac{1}{2}}2^{\frac{1}{2}}(2\pi mkT/h^2)^{\frac{3}{4}} \exp (-E_i/2kT). \qquad (60)$$

The Fermi level now lies halfway between E_i and the bottom of the conduction band (Fig. 9) if $n_e \ll N$.

Surface States

So far the discussion has centered on the interior of the material. Let us consider next the effect of adsorption. It is very likely that ionic binding will occur initially for the same reasons that lead to ionization of impurity atoms. The electrons or holes thereby concentrated on the surface will be supplied by donor or acceptor impurity atoms. This results in the formation of an electric double layer, with a sharply defined surface charge and a volume charge of opposite sign, decaying slowly toward the interior (Fig. 10). Since the potential of this layer is such that charge carriers must do work in approaching the surface from the interior, this can be considered the analog of the work-function increment in adsorption on metals. Over most of the double layer the potential V exceeds kT, so that the region is almost free of carriers (hence it is called the exhaustion layer). The charge is therefore almost entirely composed of ions, so that

$$\frac{d^2\phi}{dx^2} = \frac{4\pi Ne}{K}, \qquad (61)$$

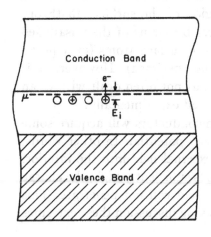

Fig. 9. Diagram of n-type semiconductor: μ, Fermi level; E_i, ionization energy of donors; O, neutral donors; $\oplus$, positive donors.

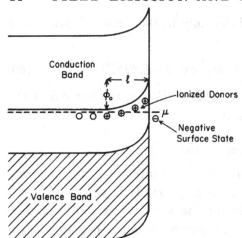

Fig. 10. Schematic diagram of energy levels near the surface of an *n*-type semiconductor with surface acceptor states.

where N is the density of impurity ions in the exhaustion layer. If N is assumed to be constant, the solution of Eq. (61) is

$$\phi_0 = \frac{2\pi N e}{K} l^2, \tag{62}$$

or

$$l = (K\phi_0/2\pi N e)^{\frac{1}{2}}, \tag{63}$$

where l is the layer thickness and ϕ_0 the contact potential due to adsorption. In most semiconductors l has values of the order of 10^{-6} cm or more.

The preceding discussion has not been specific about the adsorbate. In fact it would have sufficed to postulate the existence of electron states within the forbidden energy region of the surface. In theory these should exist even on clean surfaces by virtue of the unsatisfied valence bonds of the semiconductor's own surface atoms (or, equivalently, because of the finite size of the specimen). However, it is virtually certain that work in the past has not often dealt with clean surfaces, so that the question is not settled experimentally.

In the absence of surface states, semiconductors will acquire some volume charge on being raised to a given potential. In other words, there will be some penetration of the external field into the interior. If the density of available surface states is sufficiently high, the charge can be concentrated there, so that the field will not penetrate. This simulates metallic behavior. In the following discussion these extremes will have to be treated separately.

Emission from the Valence Band

When $kT \ll E_g/2$ and $kT \ll E_i/2$, the conduction band will be empty. Under these conditions field emission can still occur if the applied field is high enough to permit tunneling from the top of the valence band. Since each emitted electron leaves a positive hole, there will be enough conduction to balance the emission current. The latter will be given by the Fowler-Nordheim equation with an effective work function

$$\phi_{\text{valence}} = \phi + E_g, \tag{64}$$

where ϕ is the energy required to ionize an electron from the bottom of the conduction band.

Since the applied field will consequently be high, the image correction can become appreciable. The latter is given by the Nordheim factor α, Eq. (20), but with an argument

$$y = \left(\frac{K-1}{K+1}\right)^{\frac{1}{2}} 3.6 \times 10^{-4} \frac{F^{\frac{1}{2}}}{\phi + E_g}, \tag{65}$$

since the image force is changed by $(K-1)/(K+1)$ in a dielectric.

Field Emission from the Conduction Band

I. *Field Penetration Ignored.* This assumption is not realistic but it is nevertheless instructive. The population of electrons in the conduction band is assumed to be small, so that Boltzmann-statistics apply. Under these conditions the results of ordinary gas kinetic theory are valid. The rate at which electrons strike the surface from the interior is then $n(kT/2\pi m)^{\frac{1}{2}}$ electrons/cm² sec, so that the emission current is

$$i = 2nekT/(\pi m\phi)^{\frac{1}{2}} \exp\left(-6.8 \times 10^{7}\phi^{\frac{3}{2}}/F\right), \tag{66}$$

where n is the number of electrons per cubic centimeter in the conduction band and ϕ is its width (Fig. 11). The average kinetic energy of electrons in the emission direction has been taken as $\frac{1}{2}kT \ll \phi$. The number n is given by Eq. (57) or (60) and therefore the total current is temperature sensitive.

II. *Field Penetration Assumed.* If there are no surface states capable of screening the interior, the field will penetrate into the material as

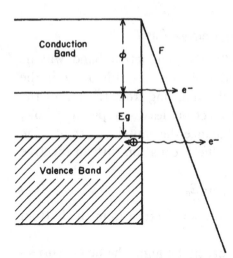

Fig. 11. Diagram illustrating field emission from a semiconductor; F, applied field with field penetration ignored. Emission from the valence band results in creation of a positive hole there.

shown in Fig. 12. Physically this corresponds to the existence of a distributed, excess volume charge of electrons near the surface and a neutralization of the normally ionized donor impurities in this region. If the conduction band is deformed by an energy V_0 not sufficient to bring it below the Fermi level, Boltzmann statistics will still apply and the emission will be given by Eq. (66) multiplied by a factor $e^{-V_0/kT}$ representing the increased electron concentration at the bottom of the band. The work function is the same as that for case I, since the height of the barrier relative to the bottom of the conduction band at the surface is unchanged. This situation may arise when the material is an intrinsic semiconductor.

Since the Fermi level is usually close to the (undeformed) conduction band when there are donor impurities and since the applied field is high, the bottom of the band will generally dip below μ in n-type material. When this happens, a "pool" of electrons will collect in this depression (Fig. 12). However, the density of filled states is now sufficient to make these electrons degenerate, so that they obey Fermi statistics. It is apparent that the highest filled level of this "pool" must coincide with the Fermi energy, at least when T is low, in order to insure equalization of the chemical potential throughout the material. Consequently the work function appearing in the Fowler-Nordheim exponent will be decreased by $\mu - V_0$:

$$\phi_{\text{effective}} = \phi - (\mu - V_0), \tag{67}$$

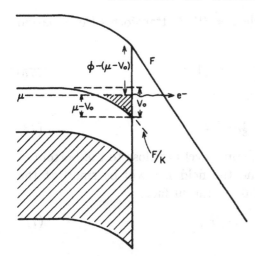

Fig. 12. Field emission from a semiconductor with field penetration: μ, Fermi level; ϕ, ionization energy from bottom of undeformed conduction band; V_0, lowering of conduction band at surface, because of field penetration; $\phi - (\mu - V_0)$, effective work function for field emission; F, applied field; F/K, field in semiconductor at the surface.

where the quantities are measured from the bottom of the undeformed conduction band (Fig. 12).

It is necessary to evaluate $\mu - V_0$. If Fermi statistics are valid, that is, when $\mu - V$ is positive, the chemical potential is given by

$$\mu = cn^{\frac{2}{3}} + V, \qquad (68)$$

where

$$c = \frac{h^2}{8m}(3/\pi)^{\frac{2}{3}}, \qquad (69)$$

so that the concentration of electrons is given by

$$n = (\mu - V)^{\frac{3}{2}}c^{-\frac{3}{2}}. \qquad (70)$$

Since the electrical potential is $-V/e$, the Poisson equation becomes

$$-\frac{d^2V}{dx^2} = \frac{4\pi ne^2}{K}, \qquad (71)$$

which can be written

$$\frac{d^2U}{dx^2} = \frac{4\pi e^2 U^{\frac{3}{2}}}{Kc^{\frac{3}{2}}} = aU^{\frac{3}{2}}, \qquad (72)$$

where

$$U = \mu - V \qquad (73)$$

and

$$a = \frac{4\pi e^2}{Kc^{\frac{3}{2}}}. \qquad (74)$$

Introduction of the new variable $p = dV/dx$ transforms the equation to

$$p \frac{dp}{dU} = aU^{\frac{3}{2}}, \tag{75}$$

which has the solution

$$p^2 = \tfrac{4}{5}aU^{\frac{5}{2}} + b, \tag{76}$$

where b is a constant. If the Fermi level lies close to the bottom of the undeformed conduction band the field p/e will be small at the point where $U = 0$, so that $b \cong 0$. At the surface,

$$p_0 = eF/K, \tag{77}$$

so that

$$F^2e^2/K^2 = \tfrac{4}{5}aU_0^{\frac{5}{2}}, \tag{78}$$

or

$$U_0 = \mu - V_0 = \nu F^{\frac{4}{5}}, \tag{79}$$

where

$$\nu = 4.5 \times 10^{-7}K^{-\frac{2}{5}} \tag{80}$$

for F in volts per centimeter and energies in electron volts. The quantity $\nu F^{\frac{4}{5}}$ must therefore be subtracted from ϕ in the Fowler-Nordheim equation, which applies otherwise unchanged (except for minor corrections) since the electron pool from which emission comes is "metallic." If the semiconductor is intrinsic, the emission will be temperature independent only as long as the population in the conduction band is adequate to replenish the metallic pool at the surface, unless the deformation takes the bottom of the pool below the top of the valence band. If there are donor impurities, they will be able to replenish the pool from which emission occurs even at low temperature.

III. *Screening by Surface States*. If there are surface states present, tunneling electrons must first overcome the internal barrier in the exhaustion layer near the surface so that emission will be given by Eq. (66) multiplied by a factor $\exp(-\phi_0/kT)$, where ϕ_0 would be given by Eq. (62) in the absence of the external field. The latter will reduce this barrier somewhat and finally break it down completely when all the states are filled with electrons so that the surface cannot accommodate more charge.

When case III occurs there will be two stable emission regions, one corresponding to (almost) perfect screening governed by the modified Eq. (66), and the other corresponding to complete field penetration governed by the Fowler-Nordheim equation, modified by Eq. (79). Between these there will be a transition region where the current rises very rapidly with applied field.

The preceding discussion indicates that field emission from semiconductors will be temperature independent either if it comes from the valence band or if there is sufficient field penetration to lower the conduction band at the surface below the Fermi level. In all other cases there will be an exponential temperature dependence. It may be possible to decide on the basis of its magnitude what the mechanism in any given case is.

This brief sketch has not mentioned many of the points discussed for metals. Thus there will be work-function anisotropies with crystallographic orientation just as with metals. These arise from the surface charge distribution and, unlike the metallic case, also decay slowly into the *interior* of the material. In other words, the work function increments appear as barriers in the bands.

EXPERIMENTAL VERIFICATION OF THE THEORY OF FIELD EMISSION

In concluding this chapter it is fitting to indicate the extent to which the theory of field emission has been vindicated by experiment. Since the information available for semiconductors is as yet very meager and incomplete, the following discussion is confined to metals. Present indications are that the theory outlined for semiconductors will also prove correct in its major features.

Energy Distribution

It has already been pointed out that the experimental work of Young and Müller[12,13] on the total energy distribution constitutes the most fine-grained proof available of the validity of the free-electron approximation in evaluating the supply function. However, the total energy distribution does not constitute too critical a test of the barrier-penetration part of the theory. This would be subjected to a more sensitive test by a comparison of theoretical and experimental normal distributions. Unfortunately, the latter are not available and not likely to become so soon.

Field-Emission Equation

A reasonable measure of the validity of the theory is afforded by a direct test of the Fowler-Nordheim equation in terms of the parameters, ϕ and F.

Dependence on F. The ordinary Fowler-Nordheim equation (51) can be written as

$$i/V^2 = a \exp\left(-b\phi^{\frac{3}{2}}/cV\right), \tag{81}$$

where a is a constant and

$$F = cV. \tag{82}$$

The general functional dependence on V can be established from graphs of $\ln i/V^2$ versus $1/V$ over ranges small enough to keep α constant. The linearity of the resultant graphs has been amply established by every worker in this field.

A more quantitative test, based on absolute magnitudes of F and ϕ has been carried out by Dyke and Dolan [14] over a very wide range of i. Their results, shown in Fig. 13, indicate that the Fowler-Nordheim equation is quantitatively obeyed within the experimental uncertainty in F (15–20 percent) if the image potential is taken into account. This also indicates that the latter is a good approximation for the potential of an electron near a metal surface up to very small distances.

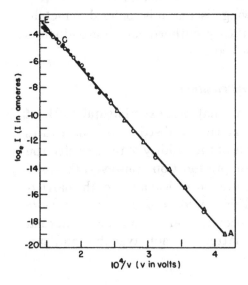

Fig. 13. Experimental and theoretical current-voltage relations in field emission, from Dyke and Dolan.[14] Curve *ACE* represents the theoretical Fowler-Nordheim curve with image corrections in the exponent only Experimental points: $\triangle$, $\bigcirc$, *DC*; $\bullet$, pulses. The deviation from the theoretical curve in the region *CE* is due to space-charge effects.

Dyke and Dolan [14] have also tested the field-emission equations in the T-F region and again find good agreement in the range where space charge is unimportant.

Dependence on Work Function. The slope of curves of ln i/V^2 versus $1/V$ is proportional to $\phi^{\frac{3}{2}}$, so that it is possible to determine the latter absolutely if the field is known, or relatively if the work function of some portion of the emitter is known. Müller [1] has obtained the relative work functions of various crystal faces of tungsten in this way in excellent agreement with thermionic determinations (except for the 110 face, whose high work function and low emission throw the thermionic measurements into some doubt).

It is also possible to obtain the work-function changes corresponding to adsorption on a field emitter from the ratio of the slopes of Fowler-Nordheim curves, if the mean work function of the clean emitter is known. This method will be discussed in more detail in Chapter 2. It suffices to state here that the contact potentials obtained in this way agree with other measurements wherever data for comparison exist. This constitutes another proof of the $\phi^{\frac{3}{2}}$ dependence predicted by the Fowler-Nordheim equation.

The foregoing indicates that the theory of field emission from metals has been subjected to fairly extensive verification and may be considered well established on experimental as well as on theoretical grounds.

CHAPTER 2

Field-Emission Microscopy and Related Topics

The field-emission microscope is the principal application of cold emission and forms the main subject of this chapter. Most other applications utilize similar emitters and techniques, so that the discussion of the former is equally pertinent.

THE FIELD-EMISSION MICROSCOPE

Field Emitters

The last chapter has shown that fields of the order of $3 - 7 \times 10^7$ v/cm are necessary for appreciable field emission. It is clearly not feasible or desirable to obtain such fields by brute force. Instead use is made of the increase in surface field at regions of high curvature, and most practical emitters consist of fine wires etched to a sharp tip, or of vapor-grown metal whiskers of suitable radius. In both cases the tip acquires a smooth, almost hemispheroidal shape after suitable heating in vacuum, in an attempt to lower its surface area and energy.

The field F at the surface of a free sphere of radius r and potential V is

$$F = V/r. \tag{1}$$

At the surface of an actual tip the field is reduced from this value by the presence of the cylindrical or conical shank, but is given to a very good approximation by

$$F = V/kr, \tag{2}$$

where $k \sim 5$ near the apex and increases with polar angle. Equation (2) indicates that adequate fields can be obtained with potentials of the order of 2–5 kv for tips of 1000-A radius. It is often easy to obtain even higher curvatures.

A detailed discussion of tip shape and field distribution cannot precede that of the emission microscope and is therefore deferred to p. 45.

Mechanism of Image Formation

The method of obtaining high fields just described leads almost automatically to the field-emission microscope, invented by E. W. Müller [1] in 1937. This can be seen by considering the trajectories of field-emitted electrons. As Fig. 3 indicates, electrons that have just emerged from the barrier have very little kinetic energy and therefore follow lines of force, at least initially. Since the emitter is conducting and hence an equipotential surface, lines of force are orthogonal to it and consequently diverge radially outward from the tip. If the emitter is surrounded by a conducting fluorescent screen as anode (Fig. 14), the diverging electrons will produce there a very much magnified image, or emission map, of the tip, as shown in Fig. 15.

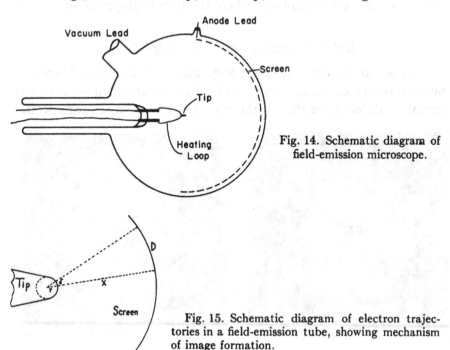

Fig. 14. Schematic diagram of field-emission microscope.

Fig. 15. Schematic diagram of electron trajectories in a field-emission tube, showing mechanism of image formation.

Ideally, the magnification would be given by x/r, where r is the tip radius and x the tip-to-screen distance. However, the emitter shank not only decreases the field at the tip but also compresses the lines of force originating there toward the longitudinal axis, so that the actual magnification is given by $x/\beta r$, where the compression factor $\beta \sim 1.5$. The resultant deformation of the image is axially symmetric and almost uniform over the visible portion of the emitter, so that almost uniform linear magnifications of the order of 10^5–10^6 can be obtained.

The tip is usually made from a polycrystalline wire by some kind of etching procedure. Nevertheless, it is almost invariably part of a single crystal because of its small size relative to the wire grains. Its spheroidal surface exposes flat facets of particularly low surface energy, blending smoothly into curved regions of continuously varying index. Since ϕ depends on orientation, the pattern shows an emission anisotropy corresponding to the crystal symmetry. Typical patterns from clean metal tips are shown in Figs. 16–19. These fade rapidly at the outer periphery, where the field is dropping off.

Field Enhancement and Local Magnification

Emission anisotropies arise not only from work-function differences but also from local variations in field. These occur whenever the local curvature differs from that of the main tip, for instance owing to the

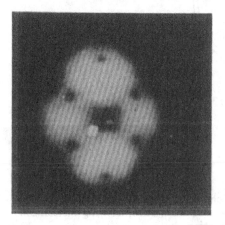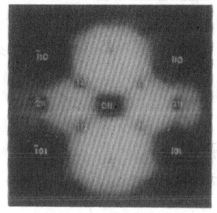

Fig. 16. Field-emission pattern of clean tungsten: (a) with two adsorbed zinc phthalocyanine molecules (small bright patterns); (b) indexed crystallographically.

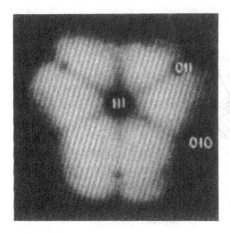

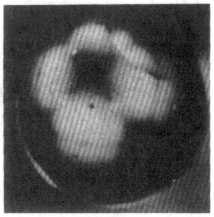

Fig. 17. Pattern of 111 oriented nickel.

Fig. 18. Pattern of 100 oriented nickel, showing 111 twin boundary in upper right-hand corner.

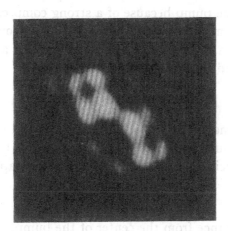

Fig. 19. Pattern of 110 oriented aluminum.

presence of protuberances or asperities. If conducting, these distort and compress the equipotentials in their vicinity, as shown in Fig. 20. This causes local field enhancement and increased emission. The distortion also leads to a divergence of the lines of force, and is equivalent to a lens effect, resulting in higher local magnification. One may think of the situation qualitatively by regarding the conducting bump as a small tip superimposed on the principal one. The resultant local

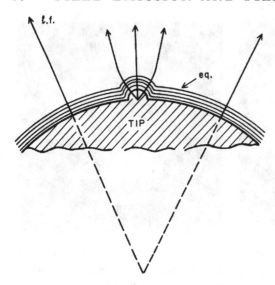

Fig. 20. Schematic diagram indicating lens effect of conducting bumps on a tip surface: *eq*, equipotentials; *l.f.*, lines of force.

magnification will exceed that of the main tip but will not correspond to the radius of the bump because of a strong compression factor.

The additional magnification M' to be expected from a hemispherical bump of radius ρ (or a ridge of that height) on a tip of radius r has been calculated by Rose[2] to be approximately

$$M'/M = 1.1(\rho/r)^{\frac{1}{2}}, \tag{3}$$

where M is the magnification of the main image. The field at the surface of such a bump can readily be shown to be 3 times that of the substrate in its vicinity, but decays with distance approximately as

$$F_{\text{bump}} = F_{\text{tip}}[1 + 2(\rho/x)^2], \tag{4}$$

where x is the distance from the center of the bump, so that the field enhancement effective for emission lies between 1 and 3.

Surface irregularities of very small size will show up as bright magnified regions if the local work function and field compare favorably with the surroundings, that is, if

$$(\phi^{\frac{3}{2}}/F)_{\text{bump}} \lessgtr (\phi^{\frac{3}{2}}/F)_{\text{tip}}. \tag{5}$$

This condition is often met by individual molecules, or small clusters,[1,3,4] and by oriented overgrowths a few atom layers in thickness or width.[5]

RESOLUTION

It is interesting to consider the factors determining resolution in the field-emission microscope.[6] Under most conditions this is limited by the statistical distribution of momenta transverse to the emission direction, rather than by diffraction effects. The reason is that the latter depend on the average, not the initial, electron wavelength. Since electrons are accelerated to almost terminal velocity in a very small distance from the tip, the average wavelength is almost identical with that at the screen and is a fraction of an angstrom.

The average tranverse electron velocity v_{yz} is proportional to $E_{yz}^{\frac{1}{2}}$, where E_{yz} is the average energy in the degrees of freedom orthogonal to that corresponding to emission. The energy E_{yz} depends on E_x, the energy in the emission direction, and at low values of T can have a maximum value of $\Delta = (\mu - E_x)$. Since barrier penetration is a sensitive function of E_x, the latter must be large so that E_{yz} will be small. Increases in field thin the barrier and permit electrons with lower E_x to tunnel. One might conclude that the consequent increase in E_{yz} would lead to a decrease in resolution. However, in tubes of ordinary construction increases in F are achieved by proportional increases in V. The time of flight of electrons is proportional to $V^{-\frac{1}{2}}$. It turns out that $E_{yz}^{\frac{1}{2}}$ varies as $V^{\frac{1}{2}}$, so that the two factors effectively cancel each other. The resulting resolution will be shown to be of the order of 20 A.

After this preliminary discussion, we shall present some simple calculations. We start by finding the displacement at the screen of an electron with transverse velocity v_t. Since the system may be replaced, for these purposes, by a concentric spherical condenser, the force acting on the electron can be taken to be proportional to r^{-2}, that is, the central-field approximation holds, so that angular momentum p_ω is conserved. This momentum is

$$p_\omega = \frac{d\theta}{dt} mr^2, \tag{6}$$

where r is the distance from the center of curvature of the tip and θ the angle between r and the radius vector (Fig. 21) describing the

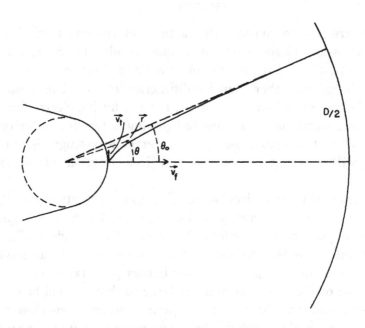

Fig. 21. Trajectory of field-emitted electron with transverse velocity component: v_f, forward velocity along flight direction if there were no transverse component v_t; θ, angle of radius vector with v_f; r, radius vector. The solid line is the electron trajectory; $D/2$ is the displacement from v_f at the screen.

flight direction of the electron if it had no transverse velocity. At $r = r_t$, the tip radius,

$$p_\omega = mr_t v_t, \tag{7}$$

so that combination of Eqs. (6) and (7) gives

$$\frac{d\theta}{dt} = \frac{v_t r_t}{r^2}. \tag{8}$$

Also,

$$\frac{dr}{dt} = \left(\frac{2e}{m}[V_0 - V(r)]\right)^{\frac{1}{2}}, \tag{9}$$

where eV_0 is the total and $eV(r)$ the potential energy, since the centrifugal correction to the latter is negligible. The electric potential $V(r)$ is given by

$$V(r) = V_0(r_t/r), \tag{10}$$

since the anode radius is very much larger than that of the tip. Consequently,

$$\frac{dr}{dt} = \left(\frac{2e}{m} V_0\right)^{\frac{1}{2}}\left(1 - \frac{r_t}{r}\right)^{\frac{1}{2}}, \tag{11}$$

so that, dividing Eq. (8) by Eq. (11),

$$\frac{d\theta}{dr} = \frac{v_t r_t}{(2eV_0/m)^{\frac{1}{2}}} \frac{1}{r^2(1 - r_t/r)^{\frac{1}{2}}}. \tag{12}$$

Introduction of the variable $u = 1/r$ permits integration of Eq. (12) and results in

$$\theta_0 = 2v_t(2eV_0m)^{-\frac{1}{2}}. \tag{13}$$

The transit time of an electron from the tip to the screen is given by

$$t = x(2eV_0/m)^{-\frac{1}{2}}[1 + (r_t/2x) \ln (x/r_t)] \cong x(2eV_0/m)^{-\frac{1}{2}}, \tag{14}$$

where x is the tip-to-screen distance. This time t is almost exactly equal to the time required by electrons of full energy eV_0 to traverse the distance x. This is due to the fact that $r_t \ll x$ and that electrons pick up almost their full kinetic energy in a distance of a few tip radii. Consequently, the transverse displacement of the electron at the screen, $D/2$, given by $\theta_0 x$, is

$$D/2 = 2v_t x(2eV_0/m)^{-\frac{1}{2}} = 2v_t t. \tag{15}$$

At first glance this result seems too large by a factor of 2. In fact, most of the literature (including the author's work) has assumed a value of $v_t t$. The reason for the extra displacement is the fact that v_t carries the electron into a region where the electric field has itself a component normal to the original flight direction.

The half-width of a region on the tip, estimated by an observer looking at the screen, will therefore be increased by an amount

$$\delta/2 = (r_t \beta/x)D/2$$
$$= 2v_t r_t \beta(2eV_0/m)^{-\frac{1}{2}}, \tag{16}$$

so that the resolution is given by

$$\delta = 4v_t r_t \beta(m/2eV_0)^{\frac{1}{2}}. \tag{17}$$

In terms of the energy associated with v_t,

$$E_t = \tfrac{1}{2}mv_t^2, \tag{18}$$

this becomes

$$\delta = 4r_t \beta \sqrt{E_t/eV_0}. \tag{19}$$

Diffraction Limit on Resolution

We consider first the limit on the resolution due to the wave nature of the electron by a method utilizing the Heisenberg principle.[6c] Electrons emerging from a region of specified width δ_0 on the tip must have a minimum transverse velocity v_t given by

$$v_t = \hbar/2m\delta_0, \tag{20}$$

since their localization within δ_0 amounts to putting a bound on the uncertainty in their position. Combination with Eq. (17) shows that

$$\delta = \frac{r_t\beta}{\pi\delta_0}\left(\frac{h^2}{2meV_0}\right)^{\frac{1}{2}} = \frac{r_t\beta\lambda}{\pi\delta_0} \tag{21}$$

since the de Broglie wavelength λ of an electron of energy V_0e is

$$\lambda = (h^2/2meV_0)^{\frac{1}{2}} = 12.5V_0^{-\frac{1}{2}}\ \text{A}. \tag{22}$$

This result could have been arrived at directly from diffraction considerations. The connection can be seen even more clearly by expressing t in terms of an average transit velocity $\bar{v}$, where $t = x/\bar{v}$, so that the displacement at the screen, $D/2$, is given from Eq. (15) by

$$D/2 = x\hbar/\bar{v}m\delta_0.$$

Now $m\bar{v}$ is the average momentum in the forward direction so that the average de Broglie wavelength is $\langle\lambda\rangle = h/m\bar{v}$ and

$$D/2 = x\langle\lambda\rangle/2\pi\delta_0,$$

which results in Eq. (21) with λ replaced by $\langle\lambda\rangle$. The difference between these wavelengths is just the factor $[1 + (r_t/2x)\ln(x/r_t)]$ neglected in Eq. (14).

Equation (21) shows that two objects δ_0 in width and δ apart will just be resolved, or that the mean resolution is

$$(\delta\delta_0)^{\frac{1}{2}} = (r_t\beta\lambda/\pi)^{\frac{1}{2}} = 2\times 10^{-4}(r_t\beta)^{\frac{1}{2}}V_0^{-\frac{1}{4}}\ \text{cm}. \tag{23}$$

Thus the resolution on a tip of 1000-A radius would be 8 A at 10 kv if it were limited by diffraction.

Statistical Limitation on Resolution

We shall now consider the loss in resolution to be expected from the statistical distribution of momenta transverse to the normal emission direction.[6a,c] Only the 0° K case will be treated, since the presence of a Boltzmann tail in the normal or transverse distribution turns out to be quite unimportant here.

We start by computing the average scalar value of the velocity transverse to the emission (x) direction, v_{yz}, for an electron of specified velocity v_x in the forward direction. This is

$$\overline{v_{yz}} = \frac{\int_0^{(v_{yz})_{max}} (v_y^2 + v_z^2)^{\frac{1}{2}} N(v_x, v_y, v_z) \, dv_y \, dv_z}{N(v_x)} \tag{24}$$

and can be evaluated by transforming to polar coordinates in the yz plane:

$$\overline{v_{yz}} = \tfrac{2}{3}(2\Delta/m)^{\frac{1}{2}}, \tag{25}$$

where Eq. (1.27) has been used and

$$\Delta = \mu - E_x, \tag{26}$$

$$E_{yz} = \tfrac{1}{2} m v_{yz}^2. \tag{27}$$

In terms of the corresponding energy this result is

$$\overline{E_{yz}} = \tfrac{4}{9}\Delta. \tag{28}$$

We find next the maximum in the normal energy distribution $I(E_x)$ by differentiating Eq. (1.33a) with respect to E_x and setting the result equal to zero. We note as on p. 16 that the factor $[E_x(\phi + \Delta)]^{\frac{1}{2}}$ is effectively constant in the region of interest, so that it may be neglected in the differentiation, whose result in terms of Δ is:

$$F/\alpha = \tfrac{3}{2} b \Delta_{max}(\phi + \Delta_{max})^{\frac{1}{2}} \tag{29}$$
$$\cong \tfrac{3}{2} b \Delta_{max} \phi^{\frac{1}{2}},$$

so that

$$\Delta_{max} \cong 9.7 \times 10^{-9} F/(\alpha\phi^{\frac{1}{2}}) \text{ ev.} \tag{30}$$

Combination with Eq. (28) gives

$$\langle E_{yz} \rangle_{max} = 4.33 \times 10^{-9} F/(\alpha\phi^{\frac{1}{2}}) \text{ ev.} \tag{31}$$

where $\langle E_{yz} \rangle_{max}$ is the average value of E_{yz} corresponding to the maximum of the normal distribution. Equation (31) shows that this quantity varies linearly with field. The resolution can therefore be found by combining Eqs. (31), (19), and (2):

$$\delta = 2.62 \times 10^{-4} \beta (r_t/k\alpha\phi^{\frac{1}{2}})^{\frac{1}{2}} \text{ cm.} \tag{32}$$

As previously indicated, this result is independent of F and places a limit of $\sim$ 20 A on the resolving power of field emitters of $\sim 10^{-5}$-cm radius. Thus the statistical momentum distribution of electrons predominates over diffraction effects, except at very high voltages.

Effective Resolution

It is probably more correct to combine the result of Eq. (32) with that of Eq. (23) vectorially as Müller has done,[1] so that the over-all resolution has the form

$$\delta = 2.62 \times 10^{-4} \beta r_t^{\frac{1}{2}} \left(\frac{1.16}{\beta V^{\frac{1}{2}}} + \frac{1}{k\alpha\phi^{\frac{1}{2}}} \right)^{\frac{1}{2}} \text{ cm.} \qquad (33)$$

Equation (33) shows that δ is independent of tip-to-screen distance and varies with the square root of tip radius. That is, it depends on objective but not on ocular magnification.

The preceding considerations must be modified if emission occurs not from the smooth surface of the tip but from a small asperity. First, the effective local radius of curvature will be smaller, and second, the image-forming electrons may originate in the asperity and thus have a momentum distribution different from that of the electron gas in the tip.

MORE ABOUT FIELD EMITTERS

The preceding section has sketched the nature of field emitters and given the basic theory of the emission microscope. In this section some of the properties of field emitters will be considered in more detail.

Tip Orientation

The tip can be indexed crystallographically from the unique symmetry of the principal directions. Thus, a 4-fold axis occurs only in the 100 and a 3-fold one in the 111 direction in cubic crystals. Examination of crystal models shows which planes are close-packed in a given structure and may, therefore, be expected to have the highest work function and lowest emission. In order of decreasing atom density, these are 110, 211, 100, for bcc and 111, 100, 110, for fcc crystals. It can, therefore, be concluded that Fig. 16 represents a 110 oriented tungsten (bcc) tip while Figs. 17 and 18 show 111 and 100 oriented nickel (fcc) tips respectively. Once two or three principal directions have been identified by inspection, their angular separations on the image can be compared with the theoretical values. In this way the compression factor β can be found and the rest of the pattern indexed.

In most metals the process of wire drawing results in preferred grain orientations along the axis. These are 110 for bcc and 111 and 100 for fcc metals. Occasionally grain boundaries intersect the tip surface. They can be identified, if planar, by projecting a photograph of the pattern onto a white sphere covered by a transparent cap, with latitude circles engraved on it. If one of these can be made to coincide with the boundary, the latter is planar and its normal is given by the pole of the cap. The normal can be identified relative to other directions on the image. In this way the line shown in Fig. 18 was identified within 2° as a 111 twin boundary.

Tip Shape

It has been pointed out that emitters are more or less spheroidal caps on conical or cylindrical shanks. It is worthwhile to discuss their shape in more detail.

Examination of Figs. 16–18 shows that flat areas, if present, are not bounded by sharp edges, since these would give rise to field enhancement and would show up by locally intensified emission. On the other hand, the presence of dark areas on these patterns, corresponding to regions of low and apparently uniform emission, is a good indication that flat, close-packed, low-energy, high-work-function faces do occur.

This conclusion is also supported by the behavior of metal atoms evaporated onto tips. Figure 22 shows bright rings of evaporated Ni atoms surrounding the 111 faces of a nickel emitter. These are formed when impinging atoms from a hot source skate over the close-packed flat faces and come to rest at their edges, where the resultant build-up is responsible for the ring-shaped areas of high emission.

Similar conclusions about the shape of emitters come from field-ion pictures taken by Müller [1] and others and from electron-microscopic shadowgraphs [7] (Fig. 23). It must therefore be concluded that tips consist of spheroidal crystals, with flat facets, corresponding to low-energy faces, blending smoothly into the curved regions of the surface.

Herring [8] has shown that the equilibrium shape of crystals may in-

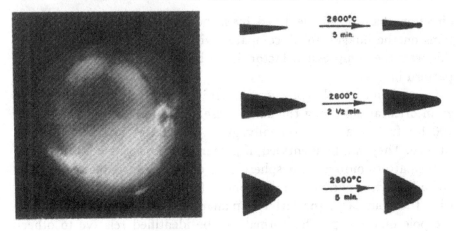

Fig. 22. Pattern resulting from the evaporation of nickel onto a nickel field emitter. The deposit shows up bright. Note bright rings around the close-packed 111 faces.

Fig. 23. Electron micrographs of field emitters before and after heating, from Dyke and Dolan.[7]

clude curved regions and that planes occur only for those orientations which correspond to point cusps in the polar surface-energy graph. The observed shapes could, therefore, correspond to frozen-in high-temperature equilibrium configurations, or to steady-state dissolution forms, since blunting occurs by atom transport from the tip when the latter is heated to high temperatures for cleaning and smoothing. Figure 23 shows forms varying from hemispheroidal caps on cones to bulbs on almost cylindrical shanks. It is readily seen that increases in tip radius beyond that of the cylindrical shank must lead to bulb formation.

The surface stress σ due to a field F at a conductor surface is

$$\sigma = \frac{1}{8\pi} (F/300)^2 \text{ dy/cm}^2. \tag{34}$$

For field emitters this amounts to $\sim 10^{10}$ dy/cm^2. There is consequently a considerable tendency for distortion by the field. This is noticed only when plastic flow or surface diffusion is rapid enough to permit rearrangement. In practice this seems to occur rapidly at $T \sim 0.3\, T_m$, where T_m (° K) is the melting point, for fields of the order of 5×10^7 v/cm. The result is facet growth with edge formation and consequent field enhancement. Figure 24 shows emission concentrated

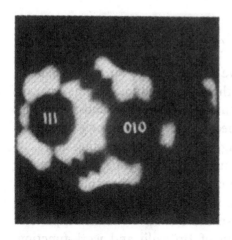

Fig. 24. Field-emission pattern from a nickel tip built up by heating with an applied field.

at the facet edges of a nickel tip built up by heating with an applied field of this magnitude.

Field Distribution at an Emitter

The details of the field distribution at an emitter depend on its shape. As the preceding section has shown, this is subject to some variation, so that accurate calculations can be made only when the former has been determined, for instance, by ordinary electron microscopy. Nevertheless it is very often adequate to use Eq. (2) with $k \sim 5$ for the field at the apex. Thus Charbonnier[9] in Dyke's group has found the empirical relation

$$k = 0.59\epsilon^{\frac{1}{3}}(x/r)^{0.13}, \tag{35}$$

where r (cm) is the tip radius, x (cm) is the tip-to-screen distance and ϵ (deg) is the emitter-cone half angle. This formula is based on an electron-microscopic examination of a large number of emitters after determination of their current-voltage characteristics, and leads to values of $k \sim 5$ for most geometries encountered in practice.

More accurate calculations have been carried out by several investigators[1,7] and again depend on the assumed tip-shank geometry. The usual procedure consists in selecting a shape that either simulates the emitter directly or generates equipotential surfaces shaped like the emitter. If the first procedure is applied to a paraboloid of revolution,

$$k = \tfrac{1}{2} \ln (x/r). \tag{36}$$

For a hyperboloid, the formula

$$k = \tfrac{1}{2} \ln (4x/r) \tag{37}$$

is obtained. Dyke and Dolan [7] have used a sphere on an orthogonal cone as skeleton and have shown that the resultant parameters can be adjusted to make an equipotential conform to almost any tip shape. They have calculated the decrease in F with angle from the apex. These results are shown in Fig. 25. The corresponding decrease in current, shown in Fig. 26, is much steeper, as one would expect from the Fowler-Nordheim equation. Consequently, the variation in F over the visible portion of the tip is necessarily small. It is this fact which makes it possible to use the Fowler-Nordheim equation in a simple way for the estimation of tip radii and work-function changes.

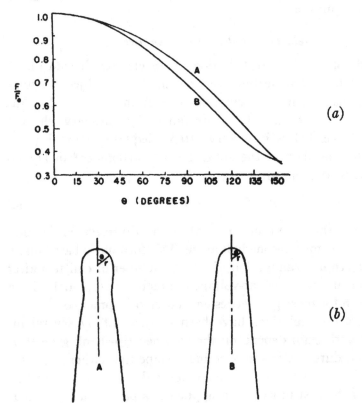

Fig. 25. (a) Field distribution at an emitter, as a function of angle from the apex for two emitters, shown in (b), after Dyke and Dolan.[7]

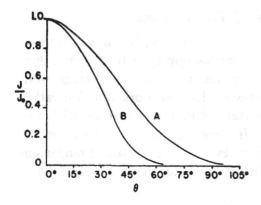

Fig. 26. Emission-current density relative to that at the apex as a function of apex angle for the emitters shown in Fig. 25b, assuming uniform work function.

USES OF THE FOWLER-NORDHEIM EQUATION

Most field-emission work requires a knowledge of emitter radius and work function or of work-function changes on adsorption. The emission equation provides a very convenient method of obtaining this information, as the following will show.

Determination of Tip Radii

The Fowler-Nordheim equation can be written

$$I/V^2 = a \exp (-b'\phi^{\frac{3}{2}}/V), \tag{38}$$

where I is the total current and

$$a = A\, 6.2 \times 10^6 (\mu/\phi)^{\frac{1}{2}} (\mu + \phi)^{-1} (\alpha kr)^{-2}, \tag{39}$$

$$b' = 6.8 \times 10^7 \alpha kr. \tag{40}$$

In these equations A is the total emitting area, α the Nordheim image-correction term, and kr the field voltage proportionality factor of Eq. (2). A graph of $\ln (I/V^2)$ versus $1/V$ is therefore linear with intercept $\ln a$ and slope $S = -b'\phi^{\frac{3}{2}}$. If α is taken to be unity and $k \cong 5$, a value of r, correct to within 20 percent, can be found at once from S if ϕ is known and uniform. A more accurate procedure consists in inserting in Eq. (2) the lower limit of kr found from S by taking $\alpha = 1$ to find a mean value of F over the voltage range of the experiment. This is used to find y from Eq. (1.21) and α from Table 1. Substitution of α in b results in an upper limit of kr, from which a new F and hence α can be found. Iteration rapidly leads to the correct value of kr, so that r can be found from Eqs. (35–37).

Determination of Work Functions

If the emitter radius is known, for instance, by direct electron-optical examination or from ion microscopy (p. 98), the procedure of the last paragraph can be inverted to obtain an average work function. It is more usual, however, that the work function rather than the radius of the clean emitter is known and that one wishes to find changes in ϕ on adsorption. In these cases kr presumably remains constant and need not always be evaluated explicitly. Comparison of the slopes of Fowler-Nordheim curves for clean and adsorbate-covered emitters yields

$$\phi_{ad} = (S_{ad}/S_{cl})^{\frac{2}{3}}(\alpha_{cl}/\alpha_{ad})^{\frac{2}{3}}\phi_{cl}, \tag{41}$$

where ϕ_{ad} and ϕ_{cl} are the average work functions with and without adsorbate and S_{ad} and S_{cl} the corresponding slopes. The factor $(\alpha_{cl}/\alpha_{ad})^{\frac{2}{3}}$ is usually so close to unity that it can be neglected.

The use of Eqs. (38) and (41) is strictly correct only if F and ϕ are constant over the emitting region, which is seldom the case. In general, the Fowler-Nordheim equation must be replaced by

$$I/V^2 = \sum_i a_i \exp\left(-b_i'\phi_i^{\frac{3}{2}}/V\right), \tag{42}$$

so that $\ln (I/V^2)$ is no longer a linear function of $1/V$. However, an average $\langle\phi\rangle$, or better $\langle b'\phi^{\frac{3}{2}}\rangle$ can be *defined* for any given V by the equation

$$I/V^2 = \bar{a} \exp\left\langle -b'\phi^{\frac{3}{2}}/V\right\rangle \tag{43}$$

and is seen to be the slope of the experimental curve of $\ln (I/V^2)$ versus $1/V$:

$$\langle b'\phi^{\frac{3}{2}}\rangle = \frac{d \ln(I/V^2)}{d(1/V)}. \tag{44}$$

Since

$$d \ln y/dx = (1/y)\, dy/dx, \tag{45}$$

Eqs. (42), (43), and (44) yield [10]

$$\langle\phi^{\frac{3}{2}}\rangle = \sum_i f_i(b_i'/\langle b'\rangle)\phi_i^{\frac{3}{2}}, \tag{46}$$

where f_i is the fraction of the total current carried by the ith emitting region. Since this quantity varies with applied field, $\langle\phi\rangle$ depends on V, in agreement with its original definition.

In practice, the curvature of ln (I/V^2) over the range of a given experiment is negligible, so that Eq. (44) yields a constant $\langle \phi \rangle$. It can be seen from Eq. (46) that this experimentally obtained value is weighted heavily in favor of the highly emitting, that is, low-work-function regions, and may in practice be taken almost identical to the lowest ϕ encountered on the tip. It is also generally permissible to neglect the variations in b' over the emitting region in calculating average values of F or r, since variations are small and Eqs. (36) and (37) are not precise enough to warrant such refinement.

However, there is one effect that must be considered in using Eq. (41) for computing average work-function changes on adsorption. If such a calculation is to have meaning, the relative emission anisotropy of the tip before and after adsorption must remain unaltered, that is, the regions contributing most to emission of the clean tip must still be the ones whose (changed) work function is measured after adsorption. Fortunately, this is generally the case, since contact-potential anisotropies are usually insufficient to alter the relative positions of the various regions.

The dangers inherent in the determination of averages can be overcome by measuring work functions of individual regions of the tip, either by photometric methods or by suitably designed current probes. In the former method the values obtained for the emitted current from high-work-function faces are almost invariably too high because of the difficulty of excluding scattered light from corresponding regions on the screen.

Changes in Preexponential Term on Adsorption

Sometimes an attempt is made to determine work-function changes by comparing currents at one voltage or the voltages required to produce a given emitted current. This amounts to assuming the constancy of a on adsorption and usually results in serious errors, since its actual changes are generally very much larger than a straightforward application of Eqs. (39) and (40) would predict.

The reason is apparently connected with the fact that adsorption causes large changes in the effective microscopic emitting area of a given region. In the case of electronegative adsorbates, emission may occur only between ad-particles. In the case of electropositive ad-atoms, the opposite may be the case, that is, the particles act like windows in the barrier. A reduction in microscopic emitting area results in either case. This argument is qualitative and physically illustrative, rather than rigorous. A quantum-mechanical treatment of such an effect has been carried out only for the case in which the perturbing potentials (that is, atoms) are distributed throughout the volume of the tunneling region (Chapter 5, p. 143), and not just the metal surface.

In addition to these effects, the depolarization of the ad-layer by the external field may produce a change in a.[11] This comes about as follows. Polarization produces a work-function change $\Delta\phi$ given by

$$\Delta\phi = 4\pi N\alpha_p F, \tag{47}$$

where N is the number of ad-particles per square centimeter and α_p their polarizability. The modification of the Fowler-Nordheim exponent takes the form

$$\exp\left[-6.8\times10^7\alpha(\phi + 4\pi N\alpha_p F)^{\frac{3}{2}}/F\right] \tag{48a}$$
$$= \exp\left[-6.8\times10^7\alpha\phi^{\frac{3}{2}}(1 + 4\pi N\alpha_p F/\phi)^{\frac{3}{2}}/F\right]$$
$$\cong \exp\left(-6.8\times10^7\alpha\phi^{\frac{3}{2}}/F\right)\exp\left[-\tfrac{3}{2}\times6.8\times10^7\alpha4\pi N\alpha_p\phi^{\frac{1}{2}}\right], \tag{48b}$$

where the F-independent term has been obtained by expansion of the last term in Eq. (48a).

Comparison of Thermionic- and Field-Emission Contact Potentials

If the dipole layer resulting from adsorption were uniformly smeared out over the surface, the contact potential would rise linearly to its final value, given by Eq. (1.2) in the surface ad-atom distance. Since the dipoles are discrete, the potential will rise more slowly, as shown in Fig. 27. In thermionic or vibrating-condenser work, only the height of the barrier is important, regardless of where the maximum occurs. In field emission, on the other hand, the slow build-up of the potential lessens the barrier for tunneling so that field-emission contact potentials should be slightly smaller than those found by other methods. The effect will be most pronounced at high fields, where the barrier is shortest, and at low coverages, where the build-up of the contact potential is slowest. Figure 28 shows the ratio of predicted to maximum contact potential as a function of coverage for several values of applied field.[12]

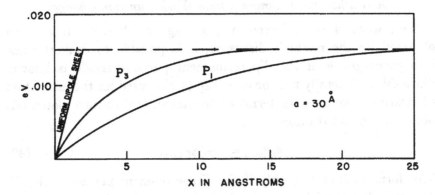

Fig. 27. Build-up of contact potential with distance from the surface for a dipole layer consisting of a discrete triangular array of dipoles; P_1 refers to the center of a unit cell and P_2 to a point close to a given dipole. The spacing between dipoles is 30 A.

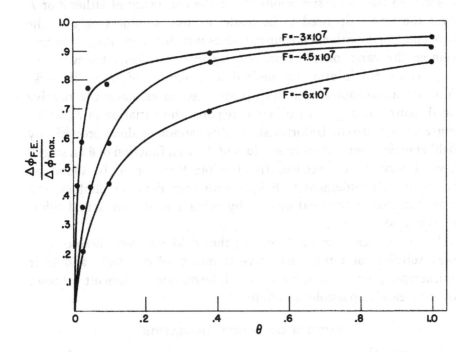

Fig. 28. Ratio of work-function increment in field emission $\Delta\phi_{FE}$ to contact potential $\Delta\phi_{max}$, as a function of coverage θ for various values of applied field; $\theta = 1$ corresponds to an ad-atom separation of 3 A.

Determinations of Coverage from Work-Function Changes

Once work-function increments, average or detailed, have been obtained by the methods discussed, it is possible to convert these into coverages, θ, if the dipole moment per ad-particle is known. Since this is generally not the case, Eq. (1.2) is used for the estimation of relative coverages, in terms of the maximum contact potentials obtainable on adsorption,

$$\theta/\theta_{max} = \Delta\phi/\Delta\phi_{max}. \tag{49}$$

This formula is correct only if the dipole moment per ad-particle P does not change with coverage. Unfortunately, reliable experimental information on this point is scanty, but the effect appears to be relatively small.

When Eq. (49) is applied individually to different regions of the crystal, so that no assumptions about the constancy of either θ or P over the whole tip need to be made, its use is subject only to the constancy of P with θ. Of course, it does not then give direct information on the variation of θ from one region to another. However, this is precisely the information desired in a number of cases. If work-function anisotropies over the tip are used to estimate the relative or absolute coverages of i different regions, the variation in P_i with i must be considered. Unfortunately, this cannot be done precisely by field emission alone. Accurate values of $\Delta\phi$ as a function of θ on single-crystal surfaces are needed. In principle these could be found by combining flash-filament technique with thermionic or photoelectric work-function determinations, or by other means, but actual information is still scarce.

It will be seen from the foregoing that field-emission microscopy is very suitable for rapid qualitative estimates of coverages and their anisotropies, but that quantitative information is difficult (though by no means impossible) to obtain.

FACTORS GOVERNING OPERATION

The operating conditions of the field-emission microscope impose fairly stringent requirements on actual emitters. These are discussed here to convey a feeling for the limitations as well as the potentialities of field-emission techniques.

Emission Currents

Visual or photographic observation of emission patterns requires current densities of $\geq 10^{-9}$ amp/cm^2 on the screen. Since tip-to-screen distances of 1–10 cm are generally used, currents of 10^{-8}–10^{-7} amp are needed. In order to obtain useful magnification and to keep the applied voltages within reasonable limits (1–20 kv), tips of 10^{-5}–10^{-4} cm radius must be used, so that minimum current densities of 10^2–10^3 amp/cm^2 on the tip are required, unless image intensification is used [13].

The narrow section of an emitter shank is usually 0.1 to 0.01 cm long. If a limit of 1 kv is (somewhat arbitrarily) set as the permissible voltage drop over this length, the emitter material must have a maximum resistivity of 3–30 ohm cm if $I = 10^{-7}$ amp. This excludes insulators and poor semiconductors.

The maximum current densities obtainable from good conductors in high vacuum are limited in practice by resistive heating of the tip and shank. If this is excessive, vaporization occurs, the evaporated atoms are overtaken and ionized by emitted electrons, and the ions are attracted back to the tip, where their presence increases the field, neutralizes the electron space charge, and leads to even more electron emission. The result is vacuum arcing and destruction of the tip.[7] With a high-melting metal like tungsten, current densities of 10^8 amp/cm^2 have been obtained with microsecond pulses [7] at repetition rates of 1000 sec^{-1}. In steady operation, currents of $\sim 10^{-4}$ amp, corresponding to a density of $\sim 10^5$ amp/cm^2, can be drawn from tungsten emitters. At the other extreme, currents of $\leq 10^{-8}$ amp have been drawn from mercury whiskers ~ 140 A in radius at 77° K.[14]

Strength Requirements

The fields required for minimum visible emission are of the order of 3–6×10^7 v/cm. For ion microscopy (Chapter 3), fields of the order of 1–3×10^8 v/cm are needed. A simple calculation shows that the stress over the circular cross section of a hemispherical emitter is given by Eq. (34), so that the material must be able to withstand stresses of the order of 10^9–10^{11} dy/cm^2 over linear dimensions of 10^{-5}–10^{-4} cm. The macroscopic strength of most metals falls far short of the theoretical value, because the presence of imperfections nucleates plastic deformation and rupture. Field emission is therefore limited to relatively strong materials unless tips are made from highly perfect crystals, such as whiskers.[15] In practice, tips can be fabricated from ordinary wires if $T_m \geqslant 1300°$ K. If whiskers are used, emission patterns can be obtained even from mercury ($T_m = 230°$ K).[12]

Attainment of Clean Surfaces

Most field-emission work requires initially clean tip surfaces, free from adsorbed layers, oxide overgrowths, and so on. The most convenient way of obtaining these conditions is to heat the tips electrically in vacuum to the point where contaminants decompose or evaporate. Since the heat of adsorption of oxygen on many metals is of the order of 5–6 ev, it is often necessary to heat tips to $> 1500°$ K to remove it. Metals with lower melting points can sometimes be cleaned by heating if their oxides are sufficiently volatile. The required temperatures often lead to excessive blunting. In practice it is almost impossible to clean metals of $T_m \leqslant 1300°$ K by heating alone.

Nonvolatile impurities dissolved in the interior are often sweated to the surface during the heating process and form oriented overgrowths there. Thus nickel containing small traces of silicon, and iron containing carbon, cannot be cleaned by heating.[5] Müller[16] has recently utilized field desorption (see Chapter 3) in conjunction with heating to clean some hitherto recalcitrant metals like iron. If subsequent experiments do not require heating to the point where more diffusion to the surface occurs, this technique can be very useful. However, the tip material must be capable of withstanding stresses of $\sim 10^{11}$ dy/cm^2, since fields of 3–5 v/A are required for the process.

Blunting

Blunting[17] by surface diffusion not only imposes a serious limitation on the amount of heating a tip may be subjected to but is also of some interest for the determination of the surface diffusion coefficient of the emitter material.[1,7] Its thermodynamic motivation arises from the fact that the chemical potential μ_s of a curved surface is higher than that of a flat one:

$$\mu_s = \mu_s° + (\gamma/r_1 + \gamma/r_2 - \sigma)\Omega, \tag{50}$$

where $\mu_s°$ is the chemical potential at zero curvature, γ the surface tension, σ the normal surface stress, if any, r_1 and r_2 are the principal radii of curvature, and Ω is the atomic volume. The surface diffusion flux J is given by

$$J = (D_s/kTA_0)\nabla\mu_s \text{ atoms/cm sec,} \tag{51}$$

where A_0 is the area per atom and D_s the surface diffusion coefficient. Diffusion occurring along the direction of the tip axis, z, across a line of length $2\pi r$, removes a shell of thickness $-dz$ and volume $-2\pi r^2 \, dz$ in time dt. Substitution in Eq. (51) yields

$$-\frac{2\pi r^2}{\Omega} \, dz = \frac{2\pi r D_s}{kTA_0} \nabla\mu_s \, dt \tag{52}$$

or

$$-\frac{dz}{dt} = \frac{D\Omega}{kTA_0 r} \nabla\mu_s. \tag{53}$$

In the present case σ is given by Eq. (34) if blunting occurs with an applied field, so that

$$\nabla\mu_s = \left[\frac{-2\gamma}{r^2} \left(\frac{\partial r}{\partial z} \right)_t - \frac{1}{8\pi(300)^2} \nabla F^2 \right] \Omega. \tag{54}$$

The factor $(\partial r/\partial z)_t$ represents the rate of change of radius of curvature with distance from the apex and can be evaluated if the tip profile is known. It is of the order of -1 to -0.5 for most emitters. An average value of ∇F^2 can be estimated from the variation of F with apex angle. Combination of Eqs. (53) and (54) results in

$$-\frac{dz}{dt} = \frac{D\Omega^2}{kTA_0 r} \left(\frac{-2\gamma}{r^2} \frac{\partial r}{\partial z} - \frac{\nabla F^2}{300^2 \times 8\pi} \right). \tag{55}$$

The contraction given by Eq. (55) can be converted into a blunting rate by multiplying by $(\partial r/\partial z)_t$. It is apparent that blunting is least for cylindrical shanks and greatest for large-angle conical shanks. Equation (55) also shows that the presence of an applied field opposes increases in r. The effect has been utilized by Dyke and Dolan[7] to retard blunting.

Vacuum Requirements

Field-emission tubes are essentially high-vacuum devices for two reasons. First, many experiments require uncontaminated emitters for long periods of time. The impingement rate of molecules from the gas phase is

$$dn/dt = P/(2\pi mkT)^{\frac{1}{2}} \text{ molecules/cm}^2 \text{ sec,} \qquad (56)$$

where P is the pressure. At $P = 10^{-7}$ mm-of-mercury and $T = 300°$ K approximately 10^{15} molecules/sec impinge on unit area. If 0.1–0.01 of these stick, a monolayer will be formed in 10–100 sec. It is apparent that pressures of 10^{-9} mm or less are needed to prevent rapid contamination.

Second, emitter life decreases rapidly with increasing pressure. Obviously, gas discharges must be avoided since demolition by ion bombardment or by vaporization from excessive emission results. Even when the pressure is low enough to avoid outright destruction, it affects emitter stability. Ion bombardment pits the tip surface, causing local depressions and elevations, with field enhancement on the latter. The resultant local increase in emission leads to more sputtering near the protuberances, etching them out even more strongly.[1] If the process is allowed to continue, the local emission soon becomes excessive and leads to vacuum arcing and destruction of the entire tip.

It was found by Dyke and Dolan[7] that the effects of sputtering and adsorption can be overcome to a large extent by operating tungsten tips at 2000° K. Since surface mobility is then very high, this requires the use of pulsed fields to prevent buildup. In this way fairly stable operation was obtained even at 10^{-7} mm-of-mercury. In general, steady operation at 10^{-6} mm-of-mercury for short periods is possible but already risky with most gases. It was first noted by Müller[1] that hydrogen is an exception, since it causes little sputtering (probably because of its low mass), so that operation at $\sim 10^{-2}$ mm-of-mercury is possible.

A SKETCH OF THE APPLICATIONS OF FIELD EMISSION

Although Chapters 4 and 5 are devoted to fairly detailed discussions of some specific applications of field emission, it may not be out of place here to present a short general survey of its actual and potential uses. A short summary of the characteristics of field emitters will first be given.

Summary of Emitter Characteristics

A field emitter is a cold cathode of small size, capable of delivering currents of 0.1–0.5 amp under pulsed and 10^{-4} amp under steady operation. Its impedance is highly nonlinear and governed mainly by the factor exp $(-6.8 \times 10^7 \phi^{\frac{3}{2}}/F)$. It represents an excellent virtual point source of electrons, $\sim$ 30 A in size, and is capable of delivering high current densities without space-charge interference. It is usually a spheroidal portion of a single crystal of known orientation and surface condition. A highly magnified ($> 10^5$) emission image with a resolution of $\sim$ 20 A can be obtained from it.

The emitter must be a good conductor, must withstand stresses of 10^{10} dy/cm^2, and should, if possible, have a high melting point, although this is not essential. Emission is sensitive to work function and hence to adsorption. Emitters can be operated at pressures as high as 10^{-7} mm-of-mercury, but are essentially high-vacuum ($\sim 10^{-9}$ mm-of-mercury) devices.

Field-Emission Microscopy

Work Functions. Emission from individual crystal faces can be measured,[18] so that work functions can be found from the Fowler-Nordheim equation, if the relation between F and V is known. This can be calculated if tip shape and dimensions are determined from electron micrographs, or can be found from the Fowler-Nordheim equation if the work function of a particular direction is known.

Surface Diffusion of Emitter Materials. Blunting rates can be measured as a function of tip temperature.[1,7] If it is possible to convert increases in tip radius into the corresponding change in length, the diffusion coefficient and its activation energy can be found, if surface diffusion is rate controlling.

This is an interesting point, incidentally, and by no means proved. In many cases it appears likely that the rate-controlling step consists of the detachment of an atom from the edge of a lattice plane (that is, partial desorption into what Stranski has called the half-lattice position) rather than of diffusion proper.

Another method is to measure as a function of temperature the rate of build-up when a field is applied,[19] or the decay without field of a built-up tip.[20] Since the initial and final points are determined from the appearance or emission of the tip and give no accurate information on the absolute material transport, only activation energies can be found in this way. Comparison of the values obtained with and without field gives some indication of its effect on the activation energy. This has been calculated theoretically by Drechsler.[21]

Surface Tension. If a field just sufficient to prevent blunting is applied, γ can be found from the balance of electrostatic and surface-tension forces,[7] by equating the right-hand member of Eq. (55) to zero.

Molecular Images.[1,3,4] Single molecules or small clusters of molecules give rise to bright spots superimposed on the main pattern. It is not completely clear at present whether this results from field enhancement and consequent extra magnification and emission or from scattering of electrons (see Chapter 5).

Surface Phases and Surface Reactions. Overgrowths of all kinds show up with additional magnification and emission. Thus the study of epitaxy on emitters is possible. Carbide,[1,22] oxide,[1,23] and silicide [5] overgrowths have been investigated by various authors. Since surface irregularities of atomic dimensions show up, field emission can detect incipient formation of second phases long before they can be picked up by macroscopic methods.

Figures 29–39 show emission patterns from nickel tips containing

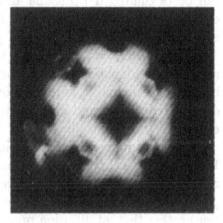

Fig. 29. Pattern from 100 oriented nickel tip, fabricated from wire containing 0.3 percent silicon, after heating to 1540° K. The silicon is in solid solution.

Fig. 30. Same tip after heating to 1470° K. Silicon is beginning to appear on the surface near 210.

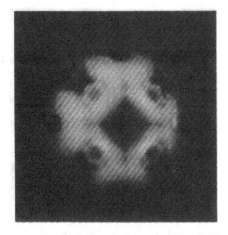

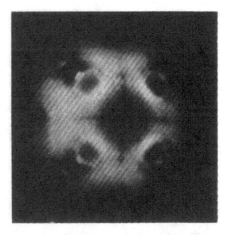

Fig. 31. Same tip after heating to 1440° K.

Fig. 32. Same tip after heating to 1400° K. Rings of silicon around 210 are beginning to form.

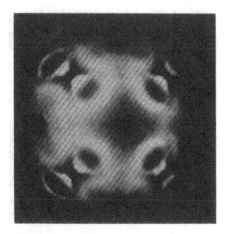

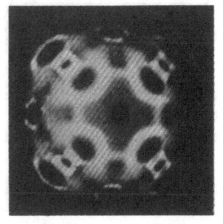

Fig. 33. Same tip after heating to 1370° K. New deposits are beginning near 110.

Fig. 34. Same tip after heating to 1325° K. New deposits are beginning to be visible near 111. Previous silicon deposits are still enlarging.

0.3 percent of dissolved silicon. This is in bulk solution only at high temperature, and begins to form a surface phase as the temperature is lowered. The patterns represent equilibrium configurations, that is, they can be approached from higher or lower temperatures. During this process the pattern goes through the equilibrium forms falling between the initial and final temperature. This suggests that the

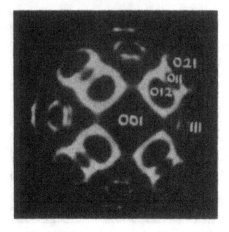

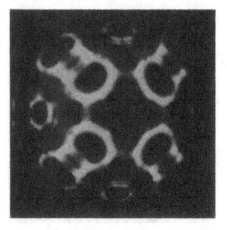

Fig. 35. Same tip after heating to 1300° K. Deposits near 111 show multiple structure.

Fig. 36. Same tip after heating to 1270° K. Field enhancement due to the silicon surface phase is becoming more marked.

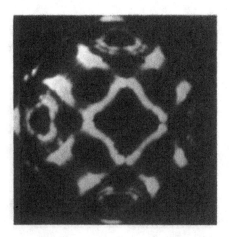

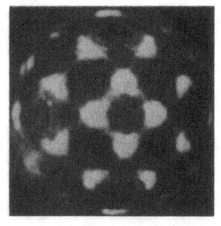

Fig. 37. Same tip after heating to 1220° K. Silicon deposits are still enlarging, with concomitant field enhancement.

Fig. 38. Same tip after heating to 1190° K. Build-up of corners is becoming marked.

actual configuration is governed by the amount of silicon present on the surface, which is determined by the temperature only after equilibrium has been reached.[5]

Oxidation and corrosion can be observed. It is difficult to estimate rates absolutely, since the observed brightness and magnification cannot be translated into accurate dimensions without a detailed

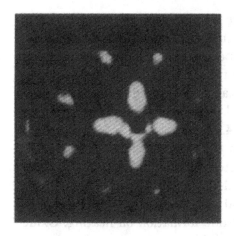

Fig. 39. Same tip after heating to 1090° K. The voltage has been reduced until only the highest points of the deposit emit, to prevent excessive currents.

knowledge of the work function and exact shape of the overgrowths. However, orientation dependence and relative rates may be estimated qualitatively. It may be possible to do more quantitative work along these lines with the field-ion microscope (see Chapter 3).

Polarizability Measurements.[1] Polarization caused by the high field existing near a negatively or positively charged tip can be used to attract a beam of atoms or molecules toward it, extending the impact zone beyond that at zero field. It is possible to estimate polarizabilities from this effect if the state of aggregation in the beam is known.

Detection Devices. Since a small number of adsorbed particles affects emission appreciably, tips can be used as detectors in molecular beams. They may also be used as integrating pressure gauges from 10^{-6} to 10^{-10} mm-of-mercury by estimating the amount of adsorption in a known time interval.

Qualitative Analysis of Gas Mixtures. Many adsorbates produce characteristic, easily recognized, pattern changes. Thus a sample of spectroscopically pure carbon monoxide was shown to contain traces of hydrogen from the fact that typical hydrogen-on-tungsten patterns with the correct mobility and desorption temperatures preceded carbon monoxide patterns when the gas was adsorbed on one side of a field emitter at low temperature.

Adsorption. Since emission depends very sensitively on work functions, even small fractions of monolayers cause perceptible changes. This, coupled with the fact that emitters are single crystals of known orientation and structure, makes the field-emission microscope ideally suited for the study of a large number of adsorption phenomena. It is possible to measure contact potentials on individual faces by comparing emission from clean and from adsorbate-covered emitters; rates of adsorption and desorption can be measured; surface diffusion can be investigated and correlated with structure. Some of these topics will be discussed in Chapter 4.

Adsorbates frequently react with the substrate when heated, and produce recognizable secondary patterns. Thus carbides having very characteristic patterns are formed from most carbon-containing compounds adsorbed on clean metal tips.[1] Similarly, the presence of oxygen in a mixture shows up by oxide formation on heating. Oxide overgrowths can also be recognized readily.

Whisker Growth.[14,15] Although the existence of metal whiskers has been known for at least five centuries, their unusual properties were not recognized until 1953.[25] Since whiskers are often very thin, it is possible to use them as field emitters without sharpening. Thus whiskers of many metals and semiconductors can be grown from the vapor *in situ* in a field-emission tube. Among other things, the growth kinetics of mercury whiskers have been followed in this way by an electrical method.[14]

Other Applications of Field Emission

There are a number of potential applications not directly connected with microscopy, utilizing the unusual electric and electron-optic properties of emitters. These are mostly in an embryonic stage of development because of the difficulties connected with stable operation over long periods of time. However, recent work, largely by Dyke and his group,[7] promises to result in a realization of these potentialities. They are sufficiently interesting to warrant mention even at this stage.

Electric Properties. The nonlinear high-voltage impedance of field emitters makes them suitable for peak sharpening, modulation, and detection in high-voltage circuits, especially in pulsed operation at microwave frequencies, where high instantaneous currents can be drawn. Thus Dyke *et al.* have built a two-cavity microwave amplifier utilizing a periodically bunched electron beam obtained by pulsing an emitter.

Electron-Optic Properties. Since emitters are virtual point sources (< 50 A) of electrons, they are useful in many lens systems. Dyke *et al.* have built a high-resolution oscillograph using an emitter as electron source. The point-source and high-voltage characteristics of emitters make them useful in x-ray devices. Marton and Schrack [26] have used an emitter as electron source in an x-ray microscope. Dyke *et al.*[7] have used a 1-in.-long comb of 40 tungsten emitters spot-welded to a wire as a line source for "flash" x-ray photography. By applying 100-kv pulses of microsecond duration they were able to draw currents of 30 amp/pulse and to photograph bullets emerging from gun muzzles, for example.

CHAPTER 3

Field Ionization and Related Phenomena

This chapter deals with the ionization of molecules by high fields [1,2a] and its applications, principally the field-ion microscope [2a] and the uses of a field-ion source in mass spectrometry.[1] The closely related subject of field desorption [3,4] and its applications will also be discussed.

FIELD IONIZATION

The first observations of field ionization were made by Müller,[5] when he admitted hydrogen at low pressure (10^{-3} mm-of-mercury) to a field-emission tube and applied a high positive voltage to the tip. A faint but highly resolved image appeared on the screen and was attributed by him to protons desorbed from the tip. He had been led to this experiment by some of his previous observations on the behavior of adsorbed electropositive metals under reversed fields,[5] which indicated a field-induced desorption.

At the time of his early hydrogen-ion observations [5] Müller believed that the improvement in resolution over electron emission resulted from the shorter de Broglie wavelength of heavy particles. Since diffraction does not limit resolution under normal circumstances, another explanation was advanced, based on the low zero-point energy of the bending modes of adsorbed H atoms.[6]

These explanations were unsatisfactory, particularly because the contact potential of H atoms adsorbed on tungsten indicated that these were covalently bound and had a slight negative charge. Mass-spectrometric experiments [1] soon justified these doubts.

In order to determine the nature of the ions produced by a field emitter, a small portion of the beam was allowed to penetrate into a mass spectrometer through a hole in the screen of an emission tube (Fig. 40). The electrical arrangement shown in Fig. 41 made it possible to keep the final energy of ions constant, regardless of generating voltages. This was important when pulsed fields were applied. In the case of hydrogen, the yield consisted mainly of H_2^+, except at very high fields ($\sim 3.10^8$ v/cm), where H^+ predominated. This result is shown in Fig. 42. With D_2 similar results were obtained at substantially identical fields. This fact immediately rules out tunneling of ions as a mechanism, since Eq. (1.9) shows this to be very sensitive to mass. It was soon found that ions could be obtained from any gas, although at varying fields. Table 2 summarizes some of the results.

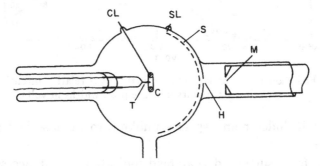

Fig. 40. Schematic diagram of field-ionization source for mass spectrometer: T, tip; C, cathode ring; CL, cathode lead; SL, screen lead; S, screen; H, aperture to mass spectrometer; M, entrance slit of mass spectrometer.

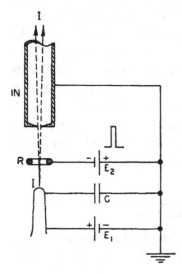

Fig. 41. Schematic circuit diagram for field-ion source: I, ion current; R, cathode ring; IN, mass spectrometer; C, capacitor, to keep tip at constant voltage relative to spectrometer in pulsed operation; E_1, final ion energy; $E_1 + E_2$, ion-generating voltage.

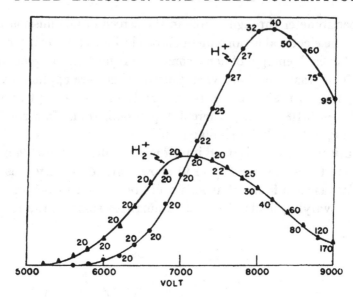

Fig. 42. Field ionization of H_2. The ordinate gives relative intensity. Figures at experimental points indicate peak widths in volts.

This list includes many species unlikely to be adsorbed as positive ions, if at all.

These facts suggested that field ionization is not necessarily connected with adsorption. This view was substantiated by the shape of the profiles of ion intensity versus energy. Figure 43 shows that these are sharp within the resolution of the machine at low fields, but develop a tail on the low-energy side when the field is increased. At very high fields the profile is smeared out over many volts and the intensity drops to zero below the original appearance energy. Since the field near the tip is very high, the potential varies rapidly in its vicinity. Thus an ion formed x angstroms from a tip at potential V has kinetic energy $V - Fx$ when it reaches the analyzer. The peak profiles indicate that ions are formed almost exclusively near the tip at low fields, but as far away as 100 A at high ones. At sufficiently high fields, ionization of all incoming particles occurs before they can reach the vicinity of the tip.

Mechanism of Field Ionization

These results make it virtually certain that field ionization is field emission in reverse, and consists in the tunneling of electrons from

TABLE 2. Ion yields from various gases with a tungsten field-ionization source.

Parent gas	Ions observed	
	primary [a]	secondary [b]
H_2	$H^+(0.5), H_2^+(0.5)$	
D_2	$D^+(0.5), D_2^+(0.5)$	
O_2	$O_2^+(1.000)$	
N_2	$N_2^+(1.000)$	
C_2H_6	$C_2H_6^+(0.8), CH_3^+$ (or $C_2H_6^{++}$)(0.2)	$C_2H_5^+, C_2H_4^+, C_2H_2^+$
C_2H_4	$C_2H_4^+(1.000)$	$C_2H_3^+, C_2H_2^+, C_2H^+$
CH_4	$CH_4^+(1.000)$	$CH_3^+, CH_2^+, CH^+, C^+$
CH_3COCH_3	$CH_3COCH_3^+(1.000)$	
CH_3OH	$CH_3OH^+(0.7),$ $CH_3O^+(0.3)$	
H_2O [c]	$H_2O^+(0.47)$ $(H_2O)_2^+(0.51)$ $(H_2O)_3^+(0.02)$ $(H_2O)_4^+(0.003)$	

[a] For the sake of clarity this table does not list isotopic peaks. Their magnitudes are in accord with accepted natural abundances.

[b] Ions listed as secondary result from collisional or vibrational breakup of primary ions as shown by pressure dependence, peak shape, and apparent fractional mass.

[c] It has been found by Beckey[8] that the ions listed here are in fact $H_3O^+ \cdot nH_2O$. See text, p. 100.

molecules into the tip.[1] Figure 44a illustrates the process for an H atom. The broken curves show a one-dimensional cut through the Coulomb potential seen by the atom's electron in the absence of external fields. A field F parallel to the plane of the paper deforms the potential as shown by the solid lines, so that the electron sees a barrier of finite height and width. If this is small enough, tunneling can occur.

Figure 44b shows a similar situation but near a metal surface. The barrier confronting the electron is smaller here than in free space, although the field is the same. This is due to a pseudo image term in the potential, arising from the interaction of the electron with the net image charge distribution induced in the metal. The latter consists of a positive charge at $-x_s$ and a negative one at $-x_n$, these

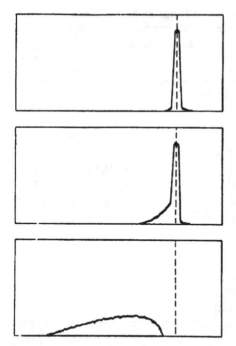

Fig. 43. Field-ion peak shapes: (*top*) low field; (*middle*) intermediate field; (*bottom*) very high field.

being respectively the distances of the electron and the nucleus from the surface. Since the nuclear motion is slow compared to that of the electron, x_n stays constant during the tunneling process. Consequently the pseudo image potential is very small when $x_e = x_n$ and approaches a true image potential as x_e decreases. It will shortly be seen that x_n has minimum values of the order of 5–8 A. At these distances V_{im} is small in any case, so that the pseudo potential is usually well approximated by Eq. (1.15).

The decrease in the barrier near the surface explains the observed peak shapes. At low fields ions are formed only where the barrier is smallest, that is, near the tip, so that sharp peaks result. At higher fields tunneling occurs even without the aid of the "image defect," so that ions can be formed farther from the surface. This produces a low-energy tail in the profile. At very high fields the ionization probability is so high that nothing reaches the vicinity of the tip, resulting in the profile of Fig. 43c.

Figure 44b shows an H atom at the distance of closest approach for which tunneling can occur with the given field. The reason for this limit is the absence of empty states below the Fermi level μ. The

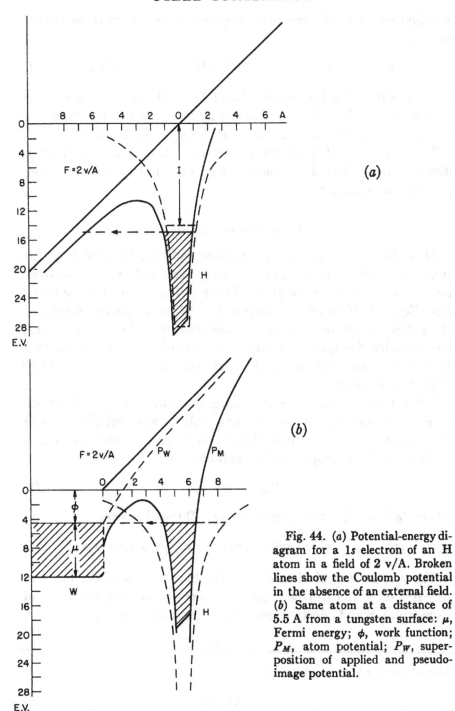

Fig. 44. (a) Potential-energy diagram for a 1s electron of an H atom in a field of 2 v/A. Broken lines show the Coulomb potential in the absence of an external field. (b) Same atom at a distance of 5.5 A from a tungsten surface: μ, Fermi energy; ϕ, work function; P_M, atom potential; P_W, superposition of applied and pseudo-image potential.

applied field must raise the tunneling electron to μ, so that the critical distance x_c is

$$x_c = (I - \phi - 3.6 \times 10^{-8}q^2/x_c)/F \cong (I - \phi)/F \text{ cm}, \qquad (1)$$

where I is the ionization potential and $3.6 \times 10^{-8}q^2/x_c$ an image term arising from the fact that the total energy of the system after tunneling includes the attraction of the ion (valence q) to the surface.

The resolution of most spectrometers is inadequate to show x_c directly. Müller has since confirmed its existence by a retardation-potential technique.[2a]

Ionization Probability

Mean lifetimes of atoms or molecules can be calculated from the tunneling probability D and the frequency with which electrons inside the molecule arrive at the barrier.[1] The former can be found from Eq. (1.9), if curves like those of Fig. 44 can be drawn. In general, the potentials are not sufficiently well known to do this accurately but an order-of-magnitude estimate is certainly possible. In drawing these curves polarization, which lowers the effective field, should be taken into account.

While D can always be estimated graphically, a simple closed expression is possible only when image effects are negligible. Under these conditions the method of Chapter 1, p. 10, applies if V_{im} is replaced by the appropriate Coulomb potential,

$$V_{coul} = Ze^2/x, \qquad (2)$$

where Ze is the effective nuclear charge; D then becomes

$$D \cong \exp\left[-6.8 \times 10^7(I^{\frac{3}{2}}/F)(1 - 7.6Z^{\frac{1}{2}}F^{\frac{1}{2}} \times 10^{-4}/I)^{\frac{1}{2}}\right]. \qquad (3)$$

The arrival rate of electrons at the barrier ν_e is of the order of 10^{15}–10^{16} sec^{-1} and can be estimated from the expectation value of momentum p_e and effective radius r_e, at least for s-states:

$$\nu_e = p_e/(2mr_e). \qquad (4)$$

The lifetime τ is then given by

$$\tau = (\nu_e D)^{-1}. \qquad (5)$$

Ion Current

Despite the relative simplicity of field ionization, the mechanism of current generation at a field emitter can be a complex function of field and temperature, and is best discussed in terms of limiting cases. Before this is done in detail, a brief qualitative description is in order.

At sufficiently high fields all particles approaching the tip become ionized before reaching it, so that the current is determined only by the supply function. As was first pointed out by Müller,[2a] this exceeds the gas-kinetic value because molecules passing near the tip are attracted to it by polarization forces.

At relatively low fields, where the total rate of ionization is small compared with the rate of arrival, the current is given by the equilibrium number of particles near the tip, divided by τ, the mean lifetime with respect to ionization. The former exceeds the zero-field value by a Boltzmann factor arising from the polarization energy of particles in the high-field region.

At fields intermediate between these extremes the situation is more complicated. To begin with, the incoming molecules have velocities in excess of the thermal values, because of polarization and dipole forces. If the atoms or molecules striking the tip are thermally accommodated there, at least in part, their rebound velocity will be less than the incoming one. Consequently they will spend more time in the high-field region on the rebound, and ionization is more likely to occur then.

If the polarization energy exceeds kT_{tip}, a fully accommodated particle is unable to leave the tip in one trajectory but will perform a series of hops, as was first pointed out by Müller.[2b] Although the capture condition is already met at room temperature for even slightly polarizable gases, adequate accommodation is not likely to occur in a single collision for particles hitting the tip with thermal plus polarization energy. However, when the tip is cooled, sufficient accommodation to prevent escape becomes more probable. Once a particle fails to escape on the first try, full accommodation to the tip temperature occurs almost certainly in the course of the subsequent hops. If the hop trajectories take particles into the ionization zone beyond x_c, the total time spent there may be sufficient to insure complete ionization of all "trapped" particles and the current will again depend only on the supply function. On the other hand, if the temperature is so low that the average hopping height is less than x_c, most incoming particles will diffuse out of the high-field region without becoming ionized. This effect was also noted by Müller.[2b]

Finally, the polarization energy may be so high in the case of polar molecules like H_2O that condensation occurs on the tip.[1] The resultant liquid or solid film usually terminates at x_c, since the ionization probability becomes extremely high there, because of the long times spent by the film molecules in the ionization zone. Under these conditions ionization of all species present in the gas phase will occur at quite low fields from the adsorbed state on the film surface and will often result in ions that are association complexes with the film molecules.[1,7]

We shall now take up the various cases in more detail.

(1) *Current Limited by Supply Function.* In order to calculate the current in this case it is necessary to find the effective capture cross section of the tip. The calculation is simplified by the fact that the polarization and dipole attractions can be approximated by central forces.[1] If the tip were a free sphere of radius r_t, the field F at a distance r from its center would be

$$F = F_0(r_t/r)^2, \tag{6}$$

where $F_0 = F(r_t)$. The field at an actual tip can be approximated by

$$F_0{}^{tip} = F_0{}^{sphere}/5, \tag{7}$$

so that Eq. (6) is still approximately valid.

The potential energy V of a particle in this field is given by

$$-V(F) = PF + \tfrac{1}{2}\alpha_p F^2, \tag{8}$$

where α_p is the polarizability and P any permanent dipole moment. If Eq. (6) holds, $V(F)$ and its derivatives are functions of r only; that is, all forces are central, so that angular momentum about the center of the tip is conserved. The problem can therefore be treated as one-dimensional if centrifugal forces are taken into account, that is, if V is replaced by

$$V' = V + p_\omega{}^2/(2mr^2), \tag{9}$$

where p_ω is the angular momentum. The latter is constant and given by

$$p_\omega = mv\rho, \tag{10}$$

where v is the velocity of the particle when very far from the tip and ρ the distance of closest approach to its center if F were zero. This is just

$$p_\omega = (3mkT)^{\frac{1}{2}}\rho, \tag{11}$$

so that

$$V' = V + \tfrac{3}{2}kT(\rho/r)^2. \tag{12}$$

The centripetal velocity, $-dr/dt$, is then

$$-drt/dt = (2/m)^{\frac{1}{2}}(E - V')^{\frac{1}{2}}, \tag{13}$$

where E is the total energy. If impingement occurs, dr/dt must be real at $r = r_t$, so that for grazing incidence

$$E - V' = 0. \tag{14}$$

The total energy E is

$$E = \tfrac{3}{2}kT \tag{15}$$

so that

$$(\rho/r_t)^2 \equiv \sigma = 1 - \tfrac{2}{3}V(F_0)/kT, \tag{16}$$

where $-V(F_0)$ is a positive quantity. Thus the effective cross section $2\pi\sigma r_t^2$ exceeds the geometric one by the factor $1 - 2V(F_0)/3kT$, which can be 10 to 100 even in the absence of permanent dipole moments, for the high field strengths encountered in field ionization.

When every particle reaching the vicinity of the tip is ionized, the current is given by

$$i = n_0 q, \tag{17}$$

where q is the ionic charge and n_0 the supply function:

$$n_0 = 2\pi r_t^2 \sigma P(2\pi mkT)^{-\frac{1}{2}}, \tag{18}$$

P (dy/cm^2) being the gas pressure.

(2) *Very Low Field.* The equilibrium concentration c_t near the tip is

$$c_t = c_g f(T_g/T_t) \exp\left[-V(F)/kT_g\right], \tag{19}$$

where c_g is the concentration in the gas phase far from the tip and T_g and T_t are the gas and tip temperatures respectively. When $T_g \neq T_t$, $f(T_g/T_t) = (T_g/T_t)^{\frac{1}{2}}$ in the immediate vicinity of the tip if full thermal accommodation occurs there and unity otherwise. In the former case, $f(T_g/T_t)$ will revert to unity in a distance slightly in excess of the average hop height. In practice this means that the factor $(T_g/T_t)^{\frac{1}{2}}$ will apply in the region chiefly contributing to the ion current at low fields.

If the total ionization does not decrease c_t appreciably, the current from a volume element $dV = 2\pi r^2 \, dr$ is

$$di = qc_t \tau^{-1} \, dV, \tag{20}$$

so that the total current is

$$i = 2\pi c_0 \int_{r_i + x_c}^{\infty} r^2 \tau^{-1} f(T_g/T_t) \exp\left[-V(r)/kT_g\right] dr$$
$$\cong 2\pi r_i^2 x_c (T_g/T_t)^{\frac{1}{2}} c_g \tau^{-1} \exp\left[-V(F_0)/kT_g\right]. \tag{21}$$

Equation (21) will be valid when the current determined from it is much less than $n_0 q$ given by Eq. (17).

(3) *Intermediate Case.* When the ionization appreciably decreases the steady-state concentration near the tip but is still far from complete, the ion current must be calculated from detailed kinetic considerations.

(a) *Single rebounds.* If the kinetic energy of particles rebounding from the tip or its vicinity exceeds the polarization energy $-V(F_0)$, they will escape completely and hopping trajectories can be neglected. Under these conditions diffusion to the tip from its shank by hopping is negligible, atoms approach the tip only from the gas phase, and the arrival rate is given by the supply function of Eq. (18).

The current generated in a shell of thickness dr is then

$$di = -q \, dn = qn \, dt/\tau \tag{22}$$
$$= qn \, dr/\dot{r}\tau,$$

where n is the number of particles entering the shell in unit time and $\dot{r}$ their radial velocity. If only particles approaching the tip are considered for the moment, integration of Eq. (22) yields for the total current on the approach

$$i_{\text{in}} = n_0 q \left[1 - \exp\int_{\infty}^{r_i + x_c} dr/\dot{r}_{\text{in}}\tau(r)\right]. \tag{23}$$

Similarly, the current produced from rebounding particles is

$$i_{\text{out}} = n_0 q \left(1 - \exp\int_{\infty}^{r_i + x_c} dr/\dot{r}_{\text{out}}\tau\right) \exp\int_{\infty}^{r_i + x_c} dr/\dot{r}_{\text{in}}\tau, \tag{24}$$

so that the total current is

$$i = n_0 q \left[1 - \exp\int_{\infty}^{r_i + x_c} \tau^{-1}(1/\dot{r}_{\text{in}} + 1/\dot{r}_{\text{out}}) \, dr\right]. \tag{25}$$

The high-field limit ($\tau \to 0$) of Eq. (25) is $n_0 q$, in agreement with Eq. (17). The low-field limit yields, by expansion of exponential terms, Eq. (21), since $f(T_g/T_t) = 1$, and the steady-state number of particles near the tip n_t is then given by

$$n_t = n_0(1/\dot{r}_{in} + 1/\dot{r}_{out})x_c. \tag{26}$$

The centripetal velocity, $\dot{r}_{in}$, occurring in Eqs. (23) to (25) is found from Eq. (13) and can be approximated by

$$\dot{r}_{in} = [\tfrac{3}{2}kT - V(F)]^{\frac{1}{2}}(2/m)^{\frac{1}{2}} \tag{27}$$

in the region of interest near the tip. At the high fields required for ionization, the polarization term $\tfrac{1}{2}\alpha_p F_0^2$ alone exceeds the thermal contribution to $\dot{r}_{in}$ even for slightly polarizable gases like He.

The rebound velocity $\dot{r}_{out}$ is more difficult to establish. If collisions were predominantly elastic, it would be equal to $\dot{r}_{in}$. However, some energy exchange is very likely, so that $\dot{r}_{out}$ will take values between $\dot{r}_{in}$ and $(kT_t/m)^{\frac{1}{2}}$ in the region of interest near the tip. A more detailed discussion of $\dot{r}_{out}$, taking polarization effects into account, will be presented in the section dealing with hopping rebounds.

As already pointed out, Eq. (25) holds only when rebounding particles escape completely from the tip. A sufficient condition for escape is that

$$-V(F_0) \leqslant \tfrac{3}{2}kT_t. \tag{28}$$

However, this is by no means a necessary condition, since complete accommodation is unlikely to occur in one collision with the tip. In the absence of sufficiently detailed theories of accommodation, it is not possible at present to predict theoretically the limits of validity for Eq. (25). Empirically it seems to hold for monatomic and homopolar diatomic species of low polarizability when, very roughly, $T_t \geqslant 200°$ K.

At the (relatively) moderate fields where Eq. (25) holds, ionization occurs principally near x_c, so that the integral in the exponent can be replaced by summation over shells of small thickness (5–10 A°), or equivalently, over small time intervals. Figure 45 shows curves of $(1 - e^{-\Delta t/\tau})$, for a hypothetical H atom gas, as a function of field, while Fig. 46 depicts its behavior with Δt at constant τ. It is clear that an increase in Δt caused by a decrease in velocity is just as effective in causing ionization as a decrease in τ. Thus most of the contribution

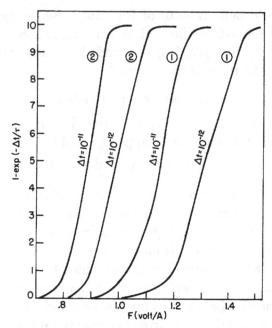

Fig. 45. The quantity $(1 - e^{-\Delta t/\tau})$ as a function of applied field for two values of Δt for a hypothetical H atom gas. Curves marked (1) refer to free atoms, curves marked (2) to atoms near a tungsten surface. The increased ionization probabilities in the latter case arise from the contribution of the pseudo-image potential to the thinning of the barrier for tunneling.

to the current in Eq. (25) will come from rebounding particles, as was first pointed out by Müller.[2a]

(b) *Multiple rebounds.* When the rebound velocity is too low for complete escape, particles will describe a series of hops until they either are ionized or diffuse out of the high-field zone. The mechanism of ionization then depends on the details of these trajectories which will now be analyzed. For simplicity it will be assumed that permanent dipole moments are absent, so that $-V(F) = \frac{1}{2}\alpha_p F^2$. It will turn out that only hops of height $h \ll r_t$ will interest us, so that centrifugal effects can be neglected. The equation of motion for a particle bouncing from the tip can then be written

$$dr/dt = \{[2E_t - \alpha_p F_0^2(r_t/r)^4]/m\}^{\frac{1}{2}}, \tag{29}$$

where E_t is the total energy along the coordinate r,

$$E_t = E - \frac{1}{2}\alpha_p F_0^2, \tag{30}$$

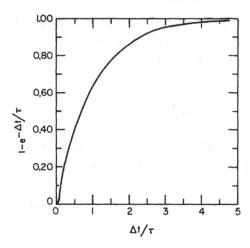

Fig. 46. The quantity $(1 - e^{-\Delta t/\tau})$ as a function of $\Delta t/\tau$.

and E is the kinetic energy along the r coordinate at $r = r_t$. The maximum value of r occurs when $dr/dt = 0$, which leads to

$$r_{max} - r_t \equiv h = \frac{E}{2\alpha_p F_0^2} r_t. \tag{31}$$

Equation (31) is similar to one derived by Müller [2b] on the basis of a somewhat different variation of F with r. The average hop height can be found at once by letting $E = \frac{1}{2}kT$. For He atoms this turns out to be 2.14 A at 20° K for $r_t = 1000$ A and $F = 3.8 \times 10^8$ v/cm, if $\alpha = 2 \times 10^{-25}$ cm³. It is interesting to note that the hop height varies linearly with tip radius.

The time required to complete a given portion of the trajectory can be calculated by integrating Eq. (29). If the variable $u = r/r_t$ is introduced,

$$t = \int_{u_1}^{u_2} a \, du/(u^{-4} - b)^{\frac{1}{2}}, \tag{32}$$

where

$$a = (mr_t^2/\alpha_p F_0^2)^{\frac{1}{2}} \tag{33}$$

and

$$b = 1 - \frac{2E}{\alpha_p F_0^2}. \tag{34}$$

Equation (32) can be rearranged to

$$t = a \int_{u_1}^{u_2} \frac{u^2 \, du}{[(1 + b^{\frac{1}{2}}u^2)(1 - b^{\frac{1}{2}}u^2)]^{\frac{1}{2}}}. \tag{35}$$

For the conditions of interest, $E/\alpha_p F_0^2 \ll 1$, so that b is very close to unity. Also, $h_{max} \ll r_t$, so that u is close to unity. The term $u^2/(1 + b^{\frac{1}{2}}u^2)^{\frac{1}{2}}$ in Eq. (35) can then be treated as a constant of value $2^{-\frac{1}{2}}$ so that

$$t = a2^{-\frac{1}{2}} \Big|_{u_1}^{u_2} \sin^{-1}(ub^{\frac{1}{4}}). \tag{36}$$

If u_2 corresponds to h_{max}, $u_2 b^{\frac{1}{4}} = 1$ while $u_1 b^{\frac{1}{4}}$ is nearly 1, so that the difference in arc sines corresponds to a very small angle. Expansion of this term then yields

$$t = a(1 - u_1 b^{\frac{1}{4}})^{\frac{1}{2}}. \tag{37}$$

If $u_1 = 1$, Eq. (37) yields the total ascent time,

$$t_1 = (m/2)^{\frac{1}{2}}(r_t E^{\frac{1}{2}}/\alpha_p F_0^2). \tag{38}$$

We must also know the time spent in the region $h \geqslant x_c$, which will be designated $2t_2$. Letting $u_1 = 1 + x_c/r_t$ we obtain from Eqs. (37) and (31)

$$t_2 \cong (m/2)^{\frac{1}{2}}(r_t(\Delta E)^{\frac{1}{2}}/\alpha_p F_0^2), \tag{39}$$

where

$$\Delta E = E(h) - E(x_c) \tag{40}$$

and $E(x_c)$ is defined by Eq. (31) as the kinetic energy along the r coordinate that would just carry the trajectory to a height x_c above the tip surface.

We are now able to consider current generation in detail. At steady state the number of atoms in the tip region, n_t, is

$$n_t = n_0(k_i + k_d)^{-1}, \tag{41}$$

where n_0 is the arrival rate and k_i and k_d are the rate constants for ionization and diffusion out of the tip region respectively. The current is then

$$i = n_0 q k_i(k_i + k_d)^{-1}. \tag{42}$$

When $k_i \gg k_d$, this reduces once again to $n_0 q$.

The diffusion rate constant is the inverse of the average time spent in the tip region by a diffusing atom (if there were no ionization). This can be found as follows. To a first approximation hopping represents a random walk, so that

$$n^{\frac{1}{2}}d = x, \tag{43}$$

where d is the width and n the number of hops. The width d is given by

$$d = 2v_t t_1, \qquad (44)$$

where v_t is the transverse velocity of hopping atoms. In order to leave the tip hemisphere, atoms must traverse some mean critical distance, of the order of r_t. The time required for this is consequently $2nt_1$, where n is to be found from Eq. (43) with $x \cong r_t$. The diffusion constant is therefore

$$k_d = 2t_1 v_t^2 / r_t^2. \qquad (45)$$

Use of Eq. (38) for t_1 with $E = \frac{1}{2}kT$ and substitution of $v_t = (2kT/m)^{\frac{1}{2}}$ leads to

$$k_d = \frac{(2kT)^{\frac{3}{2}}}{m^{\frac{1}{2}} \alpha_p F_0^2 r_t} \ \text{sec}^{-1}. \qquad (46)$$

The rate constant for ionization is given by the frequency with which atoms enter the ionization zone times the probability of ionization in a single pass, or

$$k_i(E) = \frac{t_2}{t_1} \tau^{-1} = \left(\frac{\Delta E}{E}\right)^{\frac{1}{2}} \tau^{-1} \qquad (47)$$

if $t_2 \ll \tau$. If $t_2 \geqslant \tau$, ionization will certainly occur during the diffusion lifetime of an atom and the current will be given by $n_0 q$. Equation (47) must be averaged over all energies $E \geqslant E(x_c)$, so that

$$k_i = \tau^{-1} \int_{E(x_c)}^{\infty} \left(\frac{\Delta E}{E}\right)^{\frac{1}{2}} e^{-E/kT} d(E/kT). \qquad (48)$$

On making the reasonable approximation $E^{\frac{1}{2}} \cong E(x_c)^{\frac{1}{2}}$, this yields

$$k_i = \tau^{-1} \left(\frac{kT}{E(x_c)}\right)^{\frac{1}{2}} \exp\left[-E(x_c)/kT\right]. \qquad (49)$$

If Eq. (31) is used to express $E(x_c)$ in terms of x_c and it is recalled that $x_c \cong (I - \phi)/F$, Eq. (49) becomes

$$k_i = \left(\frac{kT}{2\alpha_p}\right)^{\frac{1}{2}} \frac{r_t}{\tau(I - \phi)} \exp\left[-\frac{2(I - \phi)\alpha_p F_0}{r_t kT}\right]. \qquad (50)$$

Thus the ionization rate constant includes an activation energy that is proportional to the applied field and inversely proportional to the tip radius.

When diffusion by hopping is the only significant transport process away from the tip, the supply function n_0 can be evaluated in a way that takes implicit account of diffusion *to* the tip. At equilibrium (neglecting for the moment all ionization), the number of atoms in the tip region n_{te} is given by

$$n_{te} = 2\pi r_t^2 x_c c_t, \tag{51}$$

with c_t given by Eq. (19). If diffusion by hopping is the only process removing atoms from the tip, the supply function n_0 will be given by

$$n_0 = n_{te} k_d, \tag{52}$$

where k_d is given by Eq. (46). If it is now assumed that the occurrence of ionization does not affect the rate of arrival to first order, but only the steady-state concentration, n_0 will still be given by Eq. (52) even when $i = n_0 q$. Combination of Eqs. (51), (52), and (42) finally yields for the current

$$i = 2\pi r_t^2 x_c g c_t \frac{k_i k_d}{k_i + k_d}, \tag{53}$$

which leads to the limiting cases

$$i = n_0 q \qquad \text{for } k_i \gg k_d \tag{54}$$

and

$$i = 2\pi q r_t^2 x_c c_t k_i \qquad \text{for } k_i \ll k_d. \tag{55}$$

Equation (55) closely resembles Eq. (21) except that explicit assumptions about the ionization process have been made so that τ^{-1} has been replaced by $k_i = (t_2/t_1)\tau^{-1}$.

It is interesting to apply these results to a specific example. The cases of He atoms on a tungsten tip of 500-A radius at 20° K and 4° K will be considered at an applied field $F_0 = 3.8$ v/A. Various pertinent quantities and the results are listed in Table 3. It is seen that for the same field the current at 4° K is 0.02 of that at 20° K because of the changes in the rate constants. This is in qualitative agreement with the behavior observed experimentally by Müller [2b]. At the present time sufficient quantitative data for a rigorous test of the theory, involving all the parameters, are not available.

(4) *Film Formation.* The preceding considerations must be modified if the polarization energy is sufficient to cause condensation on the tip. In the case of H_2O, for example,[1] this occurs at room tempera-

TABLE 3. Field-ionization data for He atoms at low temperatures.[a]

Quantity	T_i (° K)	
	20	4
kT (erg)	2.76×10^{-15}	5.5×10^{-16}
$\frac{1}{2}\alpha_p F_0^2$ (erg)	3.22×10^{-13}	3.22×10^{-13}
$\langle h_{max} \rangle$ (A)	1.1	0.21
x_c (A)	5.3	5.3
$2t_1$ (sec)	2.1×10^{-12}	0.95×10^{-12}
$\langle d \rangle$ (A)	6	1.3
k_d (sec^{-1})	10^8	10^7
τ (sec)	10^{-11}	10^{-11}
k_i (sec^{-1})	7×10^9	2×10^5
k_i/k_d	70	0.02
i	$n_0 q$	$0.02\, n_0 q$

[a] Data calculated for a tungsten tip of 500-A radius and an applied field $F_0 = 3.8$ v/A. The polarizability of He is taken as 2×10^{-25} cm^3. Quantities in brackets are average values, based on thermal equilibrium. The value of τ is calculated from Eq. (3) and $\nu_e = 10^{-15}$ sec^{-1}.

ture and a gas pressure $< 10^{-4}$ mm. Equation (19) provides a rough criterion for condensation. If the pressure P calculated from it is appreciably less than the vapor pressure of the liquid P_0, it is unlikely that it will occur since Eq. (19) provides only an upper limit to P. If the possibility of condensation is indicated a more careful appraisal is necessary. The effective vapor pressure P_L of the condensed films (as if it existed at zero field and curvature) can be found by equating the rates of arrival and departure of molecules at its surface, if the flow of liquid from the emitter shank to the tip is neglected:

$$P_L = P_{gas}\left(1 - \frac{2}{3}\frac{V_G}{kT}\right) \exp\left(-\int_{x_c}^{\infty} \frac{dr}{\dot{r}_{in}\tau} - \frac{2\gamma\bar{v}}{r_t kT} + \frac{\sigma}{kT}\right), \qquad (56)$$

where $-V_G$ is the (numerically positive) field–gas-molecule interaction energy of Eq. (8), γ the surface tension of the film, $\bar{v}$ its molecular volume, and σ an energy term to be discussed presently. The three terms in the exponential factor in the right-hand member of Eq. (56) are from left to right the chance of an impinging gas molecule's reaching the liquid surface without ionization, a correction for curvature, and a correction to zero field. The energy σ occurring in the last named is given by

$$\sigma = \bar{v}(K - 1)F^2/8\pi + V_G \qquad (57)$$

if an evaporating molecule is almost certain to be ionized just outside the film. If the ionization probability is small, the term V_G must be omitted from σ if the field decays to a small value in a distance short compared to the mean free path of a gas molecule. The dielectric constant K appearing in Eq. (57) must be that appropriate at high fields.

The value of P_L calculated from Eq. (56) can be used with the appropriate isotherm, if available, to estimate the film thickness. Since isotherms on the relevant single-crystal surfaces are likely to be rare, one will generally be forced to compare P_L with the condensation pressure P_0. It can be assumed roughly that multilayer formation is likely if $P_L/P_0 \sim 0.5$. If the calculated thickness exceeds x_c, the film will nevertheless terminate there, even if the field is weaker than that required for ionization in the gas phase. The reason for this is twofold. First, the time spent by a condensed molecule in the ionization zone is effectively infinite compared to the transit time of a gas molecule, and second, its ionization potential is reduced by $1/K^2$. Since the local field within the dielectric is only decreased by $1/K$, the net effect is a large increase in ionization probability.

This argument also indicates that ionization will occur principally from molecules or atoms adsorbed on the surface of the condensed film if its thickness equals or exceeds x_c for the species being ionized. Thus the presence of condensed films can drastically reduce the field required for ionization. Indirect confirmation for this comes from the fact that Schissel [7] was able to use silver tips for field-ionization work under film conditions. Normally these tips would be expected to be much too weak for this purpose. The fact that heating can lead to the cessation of ionization unless the field is greatly increased also lends support to this view. It is also very likely that the resultant ion may be an association complex with the film species (for example, H_2O polymers or the entities found by Beckey) [8]. The latter also believes that the film may be a priori ionized at high fields in the case of polar molecules. (See Appendix 2.)

A quantitative prediction of film formation based on Eq. (56) is difficult, but it is easy to understand the situation qualitatively.

The ionization probability of incoming molecules varies so steeply with field that it may be regarded as a step function. Condensation is a cooperative phenomenon which occurs over a very narrow pressure range and can also be taken as a step function of F. The right-hand member of Eq. (56) can therefore be regarded as the product of two step functions of F of opposite sense. Layer formation will occur whenever the condensation step occurs at a lower value of F than the ionization step. The separation of the two on the F axis then gives the range over which field ionization will occur from the layer surface, if at all. It is evident that the relative position of the steps depends on the gas pressure and temperature as well as on the pertinent molecular constants. In practice, the sudden onset or cessation of condensation may be smeared out somewhat by the flow of liquid from the shank to the tip on the one hand, and on the other by the heating due to the bombardment of the film by incoming molecules, arriving with kinetic energy $-V_G$.

FIELD DESORPTION

In the course of experiments on the behavior of adsorbed layers of barium and thorium on tungsten field emitters,[5] E. W. Müller noted that high inverse fields (with respect to electron emission) resulted in permanent decreases in field emission and concluded that field desorption of Ba^+ and Th^+ ions had occurred. As already pointed out, these experiments led him to develop the ion microscope.

More recently, Müller showed that field desorption of oxygen from tungsten and of tungsten from its own lattice occurs at sufficiently high fields [3] (4–5 v/A). In all these cases the assumption is made that the desorbed species is ionic. Direct proof of ionic desorption in some cases can be obtained mass spectrometrically,[1] as described in the next section.

Mass-Spectrometric Observations

Mass spectra from a field-ion source sometimes show that one or more peaks do not broaden with field while others do. Thus the spectrum of methanol, shown in Fig. 47, has a peak corresponding to CH_3OH^+ that broadens and one corresponding to CH_3O^+ that does not. Since lack of broadening implies a fixed locus of ionization, it is very likely that this is the tip itself, and that such ions correspond to

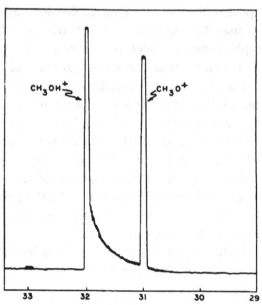

Fig. 47. Field-ion spectrum of methanol. Note the broadening of the parent peak.

adsorption products. Added weight is lent to this interpretation by the fact that the ions obtained from a given substance depend on the prior treatment of the tip. Thus it is possible to carbidize a tungsten tip by heating it to 700° K in C_2H_5OH. The spectrum then changes from a simple to a very complex one, indicating that the tip catalyzes the synthesis of hydrocarbons, water, and other substances. Further, the intensity of nonbroadening relative to broadening parent peaks increases with decreasing pressure, as one would expect in chemisorption.

Almost conclusive proof is furnished by pulsed-field experiments. At steady fields the ion current results from the balance of the following processes: (1) diffusion to the tip; (2) ionization in the gas phase; (3) adsorption; (4) (field) desorption from the ad-layer. The steady-state number of molecules contained in the ionization region is very small compared with that adsorbed on the tip at even moderate coverage, since the density of a condensed phase is much higher than that of a dilute gas. At low pressures it is possible to minimize the number of molecules diffusing into the ionization zone during a pulse if this is short. It is therefore possible to shift the contribution to the current from process (2) to (4) by increasing the time interval between pulses. When this is small, the dc situation is approximated, but as it increases the ratio (2)/(4) approaches zero. This is confirmed experimentally by the results shown in Fig. 48.

Theory of Field Desorption

The phenomena of field desorption can be treated most simply in terms of the potential-energy diagrams of the metal-adsorbate system.[4] Before doing so, field ionization will be reconsidered from this point of view.

Figure 49a shows a potential-energy diagram for a nonadsorbed atom A and a metal surface M as a function of the atom-surface separation x. This curve, marked $M + A$, is a horizontal straight line. The curve $M^- + A^+$, corresponding to the approach of A^+, has the form of an image potential [Eq. (1.10)] at moderate values of x and lies above $M + A$ by $I - \phi$ at large x, if the electron from A is in the metal. Since this charge distributes itself, its contribution to the potential is negligible at ordinary surface-to-volume ratios.

In the absence of external fields the curves intersect only at very small distances from the surface, if at all. An applied field deforms $M^- + A^+$ while merely displacing $M + A$ downward by a polarization term $\frac{1}{2}\alpha_p F^2$, so that the curves appear to intersect at x_c (Fig. 49b). However, the degeneracy implied by crossed potential curves is permissible only if there are spin or symmetry differences between the states involved. Generally, these are absent so that the curves will separate into the new states shown in Fig. 49b. The separation will usually be very small, since there will be little overlap of the atomic and ionic wave functions at x_c.

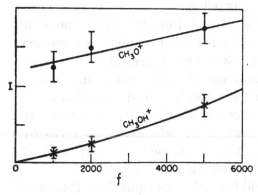

Fig. 48. Relative abundances of CH_3OH^+ and CH_3O^+ in the field-ion spectrum of methanol, as functions of the pulse repetition rate f.

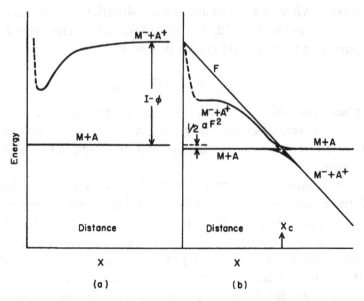

Fig. 49. Potential energy as a function of atom-surface separation x for a non-interacting atom approaching a metal surface: $M + A$, metal plus neutral atom; $M^- + A^+$, ionized atom and electron from the former in the metal; I, ionization potential; ϕ, work function; (a) zero applied field; (b) applied field F. The ionic curve is deformed strongly and intersects the atomic one at x_c, resulting in the new states shown. Broken lines correspond to a nonadiabatic transition. The quantity $\frac{1}{2}\alpha F^2$ represents the polarization energy of neutral atoms in the field F.

Adiabatic transitions (in the Ehrenfest sense) of an atom moving away from the surface will therefore lead to ionization at x_c. The particle will then follow the lower curve. Nonadiabatic transitions, corresponding to a crossing of the curves along the broken lines, occur when the atom's motion in the transition region is too rapid for electronic readjustment, that is, tunneling.

An atom approaching x_c along $M + A$ from the right represents an excited state in the presence of a field. A nonradiative transition to the ground state can occur at $x \geqslant x_c$ by field ionization.

The meaning of adiabatic and nonadiabatic transitions is particularly clear here. The former corresponds to field ionization at x_c; its rate can be found in the manner already described; x_c is precisely the minimum atom-surface separation for field ionization given by Eq. (1). The nonadiabatic transition corresponds to an insufficient rate of field ionization, that is, too small a value of $\Delta t/\tau$.

The two ways of looking at field ionization are equivalent. However, the treatment of this section brings out more clearly the necessity for energy conservation in the total system. This can best be seen by noting that the separation between the curves at any value of x is the net ionization energy of the atom, if the electron enters the metal. At x_c this quantity is zero.

This method will now be applied to field desorption.

(1) $I \gg \phi$. Figure 50a shows the situation for relatively strong covalent chemisorption, when $I - \phi$ is large so that the ionic curve does not intersect the ground state. Under these conditions ionic evaporation from the latter can occur only by a nonvibrational transition to the upper state, for instance, by photoexcitation. Vibrational transitions would be preceded by desorption along the atomic curve even if this should intersect the ionic one very near the surface.

Figure 50b shows the same system in the presence of a fairly high positive field. Except for the attractive portion of the ground-state curve, this diagram is exactly the same as that of Fig. 49 for field ionization, and the arguments apply unchanged. Vibrational desorp-

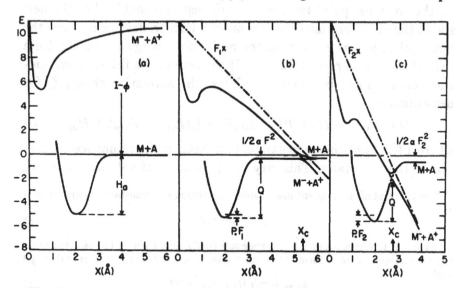

Fig. 50. Potential-energy curves for neutral and ionic adsorption at a metal surface with large separation between states: I, ionization potential; ϕ, work function; x, adsorbate–surface distance; F, applied field; $\frac{1}{2}\alpha F^2$, polarization energy of atom in field F; $P.F$, field-bond dipole interaction energy; H, heat of adsorption of atom; Q, activation energy of desorption; x_c, intersection of atomic and ionic curves; (a) zero field; (b) moderate field; (c) high field.

tion requires the zero-field activation energy H_a (except for dipole and polarization terms), but can be followed by field ionization at x_c. The rate constant k_{ion} for desorption * resulting in ions is given by

$$k_{ion} = \nu \exp\left(-\frac{H_a + PF - \frac{1}{2}\alpha_p F^2}{kT}\right)\left\{1 - \exp\left[-\left(\frac{m}{kT}\right)^{\frac{1}{2}} \int_{x_c}^{\infty} \frac{dr}{\tau(r)}\right]\right\}$$
$$\text{sec}^{-1}, \quad (58)$$

where ν is the vibrational frequency of the adsorbed particle along the desorption coordinate and PF the bond-dipole field interaction. Equation (58) is based on the assumption that the velocity of desorbing particles near x_c is thermal. The corresponding expression for atomic desorption is

$$k_{at} = \nu \exp\left(-\frac{H_a + PF}{kT}\right) \exp\left[-\left(\frac{m}{kT}\right)^{\frac{1}{2}} \int_{x_c}^{\infty} \frac{dr}{\tau(r)}\right] \text{sec}^{-1}. \quad (59)$$

The polarization term has been omitted here since the field decays in distances less than the mean free path of desorbing atoms.

Figure 50c shows the system when the applied field is so high that x_c corresponds to the attractive part of the neutral adsorption curve. At the crossover point the curves again separate and lead to the new states shown. Vibrational excitation can lead to desorption over the potential saddle at x_c, so that the activation energy is reduced from that of the zero- or low-field case. If the zero of the field-free ground-state curve, $V(x)$, is taken at infinity, the activation energy Q for desorption is

$$Q = V(x_c) + H_a + PF - \frac{1}{2}\alpha_p F^2 - \frac{1}{2}\Delta E_{res} \approx V(x_c) + H_a, \quad (60)$$

where $V(x_c)$ is of course negative, and ΔE_{res} is the resonance separation of the new states. This may be appreciable at values of x_c where

---

* In this and all subsequent discussions, desorption rate constants k_d are written as

$$k_d = s\nu e^{-Q/kT}, \quad (a)$$

where ν is a frequency and s a transmission coefficient, such as the last factor in Eq. (58). A more correct formulation on the basis of absolute reaction-rate theory [9] leads to

$$k_d = s(f^*/f_i)(kT/h)e^{-Q/kT}, \quad (b)$$

where f_i and f^* are respectively the partition functions of the ad-complex in the initial and activated state (that is, at the top of the potential barrier which lies Q v above the minimum of the potential curve), kT/h a universal frequency ($\sim 10^{13}$ sec^{-1}), and s a transmission coefficient to take care of the possibility of reflection at the top of the potential barrier. The ratio f^*/f_i will be close to unity the overlap between the electronic wave functions corresponding to the covalent and ionic states is significant, but will be neglected for the present.

The rate constant for field desorption is then

$$k_{des} = vse^{-Q/kT}, \tag{61}$$

where v is a vibration frequency (see footnote, p. 88), and Q the activation energy given by Eq. (60); s is a transmission probability, that is, the chance that a particle with sufficient energy to pass over the saddle point will in fact make the adiabatic transition and become ionized.

If the field is sufficiently weak to place the saddle point in a region where the field-free ground-state curve is only weakly attractive, the ad-particle can be treated more or less as a free atom. In that case s can be found as a field ionization probability,

$$s = (1 - e^{-t/\tau}), \tag{62}$$

where t is the time spent during one vibration in the region $x \geqslant x_c$ where ionization can occur and τ is the mean life with respect to field ionization. Strictly speaking, t is a function of vibrational energy E, so that the rate constant should be written

$$k_{des} = \frac{\int_Q^{\infty} (1 - e^{-t(E)/\tau}) v e^{-E/kT} \, dE}{\int_0^{\infty} e^{-E/kT} \, dE} \tag{63}$$

which yields

$$k_{des} = v e^{-Q/kT} - \frac{v}{kT} \int_Q^{\infty} e^{-t(E)/\tau} e^{-E/kT} \, dE. \tag{64}$$

The substitution $\epsilon = (E - Q)/kT$ in the integral of Eq. (64) leads to

$$k_{des} = v e^{-Q/kT} \left(1 - \int_0^{\infty} e^{-t(\epsilon)/\tau} \, d\epsilon \right). \tag{65}$$

The term in parentheses is the s of Eq. (61) and affects only the preexponential term of the rate constant, since it does not contain Q. The integral in Eq. (65) can be evaluated only if the shape of the potential curve is known. In general this may not be necessary, how-

---

for most cases of interest here unless the ground state corresponds to immobile and the activated state to mobile adsorption. Thus Eq. (a) can be thought of as a shorthand notation, with the understanding that v can differ from a vibrational frequency whenever entropy effects are important.

ever, since $t(E)/\tau$ may be sufficiently high at the fields required for desorption to make the integral small.

In the classical harmonic oscillator approximation $t(\epsilon)$ can be shown to be

$$t = \tau a \epsilon^{\frac{1}{2}}, \tag{66}$$

with a given by

$$a = (\pi \nu \tau)^{-1}(kT/Q)^{\frac{1}{2}}. \tag{67}$$

Integration of the last term in Eq. (65) then gives the transmission coefficient s:

$$s = \tfrac{1}{2}\pi^{\frac{1}{2}}a[1 - \text{erf } (a/2)] \exp (a^2/4), \tag{68}$$

which has the limiting values

$$\lim_{a \to \infty} s = 1, \tag{69a}$$

$$\lim_{a \to 0} s = \tfrac{1}{2}\pi^{\frac{1}{2}}a, \tag{69b}$$

so that the rate constant k_{des} is given in these limits by

$$k_{\text{des}} = \nu e^{-Q/kT}, \qquad (a > 5) \tag{70a}$$

$$k_{\text{des}} = (2\pi^{\frac{1}{2}})^{-1}(kT/Q)^{\frac{1}{2}}\tau^{-1}e^{-Q/kT}. \qquad (a < 0.1) \tag{70b}$$

At $300°$ K, $\tau a \cong 5 \times 10^{-14}$ sec if $\nu^{-1} \cong 10^{-12}$ sec, so that $s \geqslant 0.3$ if $\tau \leqslant 10^{-13}$ sec, which is probable in most cases.

When there is strong covalent binding at x_c in the absence of the field, that is, when x_c lies on the strongly attractive part of the original ground-state curve, τ loses its previous simple meaning. It is then necessary to carry out a time-dependent perturbation calculation for the transition probability. Since this involves a much more detailed knowledge of the system than is generally available, the following simplified treatment can be used.

The lifetime with respect to the adiabatic transition near the saddle point is connected by the uncertainty principle with the diffuseness in energy at the crossover region. The latter is equal to the resonance separation of the new states. Consequently,

$$\tau \cong \hbar/\Delta E_{\text{res}}, \tag{71}$$

and this value must be used in Eq. (65). Equation (71) indicates that the transmission coefficient s increases with the resonance separation of the curves, which, as was pointed out, increases with decreasing x_c, that is, with increasing field. It is therefore probable that s is large over the entire field-desorption range. (See Appendix 2.)

The mechanism just outlined corresponds to complete or partial desorption in the neutral state, followed by field ionization. When the field is strong, the ionization condition is met when the vibrational displacement corresponds to only partial desorption at x_c, that is, when

$$Fx_c = I - \phi - 3.6q^2/x_c - V(x_c). \tag{72}$$

In every case the rate constant takes the form of Eq. (61) with Q a function of the applied field. The third term on the right-hand side of Eq. (72) is correct if the ionic curve still conforms to an image potential at x_c. If the latter is small, this may not be a good approximation.

(2) $I - \phi$ small. Figure 51 shows neutral and ionic curves when $I - \phi$ is small so that crossing occurs even in the absence of applied fields. In most cases there will be no degeneracy, so that the curves will separate. The adiabatic approach of an atom will therefore result in an ionic ground state, as shown. (It is interesting to note in passing that the probability of an adiabatic transition involving electron transfer can be treated as a field-ionization problem.) Vibrational excitation can lead to direct ionic evaporation even in the absence of applied fields, if the desorption is nonadiabatic. The phenomenon is known as surface ionization.

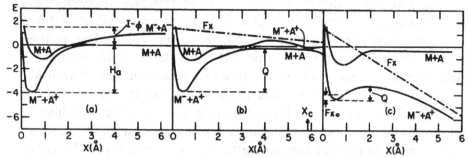

Fig. 51. Potential-energy curves for neutral and ionic adsorption at a metal surface when $I - \phi$ is small. Notation as in Fig. 50; Fx_0, increase in binding due to field; (a) zero field, in which the curves cross, leading to an ionic ground state; (b) moderate field, giving rise to a double intersection so that desorption can proceed either along the bottom curve (adiabatic) or by a doubly nonadiabatic path; (c) high field, which removes the intersection so that the ground state remains purely ionic.

Figure 51b shows the effect of a moderate positive field on this system. Ionic evaporation can occur by two paths, as indicated on the diagram. The first consists of desorption followed by field ionization and corresponds to case (1b). The second consists of direct ionic evaporation along a doubly nonadiabatic path, with an activation energy Q given by the energy difference between the minimum of the ground-state curve and the Schottky saddle:

$$Q = Q_0 - 3.8q^{\frac{3}{2}} \times 10^{-4}F^{\frac{1}{2}} + qFx_0, \tag{73}$$

where

$$Q_0 = H_a + I - \phi. \tag{74}$$

The term qFx_0 represents the field-ion interaction at the equilibrium distance x_0.

A case of greater interest is depicted in Fig. 51c. Here the applied field deforms the ionic curve so strongly that no intersection with the atomic curve occurs. The ground state is therefore purely ionic at all distances and evaporation occurs by vibrational excitation over the Schottky saddle without need for field ionization. The rate constant of the process is given by

$$k = \nu e^{-Q/kT} \text{ sec}^{-1}, \tag{75}$$

with Q given by Eq. (73). Equation (75) is formally equivalent to that derived by Müller.[3]

There is some indication that cases intermediate between (2b) and (2c) occur, where the desorption path requiring lowest activation energy is that corresponding to (2c), but where this intersects the atomic curve. Temporary trapping in the metastable atomic minimum can then occur and leads to an apparent reduction in ν. The field desorption of barium from tungsten seems to fall in this category.[4]

The field evaporation of thorium from tungsten, studied by Müller,[3] seems to provide a rather clear-cut case of mechanism (2c).[4] Müller's data are in the form of field strengths versus temperature T for constant desorption times $t = 3$ sec. Equations (73), (74), and (75) can be rearranged to give

$$F^{\frac{1}{2}} = \frac{Q_0 + qFx_0}{3.8 \times 10^{-4}q^{\frac{3}{2}}} - T\frac{\log_{10}(t\nu)}{1.9q^{\frac{1}{2}}}. \tag{76}$$

A graph of $F^{\frac{1}{2}}$ versus T should therefore be linear with a slope given by the coefficient of T and an intercept proportional to Q_0 if $qFx_0 \ll Q_0$. The data are plotted in this way in Fig. 52 for low coverages of thorium on tungsten. If $q = 1$, $\nu = 5 \times 10^{10}$ sec^{-1} from the slope of Fig. 52. If desorption occurred as Th^{+2}, ν would have to be 10^{31} sec^{-1}, which is clearly absurd. Field-desorption experiments can therefore be used to determine the charge on the desorbing ion, and hence usually on the bound ion, even without mass-spectrometric analysis. The intercept of Fig. 52 yields $Q° = 6.45$ v. Since the first ionization potential of Th is not known, H_a cannot be found from this value. However, a value $H_a = 7.7$ v was found by Langmuir.[10] If this is correct, an ionization potential of $I = 3.75$ v can be calculated. This example illustrates the usefulness of field desorption in the case of positive adsorbates.

The mechanism of field evaporation of ions from their own lattice, investigated and used by Müller[3] for the controlled piecemeal decomposition of emitters, is probably intermediate between mechanisms (1c) and (2c). That is, the normal state of an atom in the lattice may correspond to electroneutrality, but the intersection of the ionic with the ground-state curve may occur in such a way that the Schottky saddle represents the highest point on the desorption path. In that case the experimentally found activation energy is that of Eq. (71) with the term qFx_0 omitted.

THE FIELD-ION MICROSCOPE

The field-ion microscope invented by E. W. Müller[2a,5] was originally run with hydrogen, although it is now commonly used with other gases, primarily helium. In view of the preceding sections, the mechanism of image formation is fairly obvious. Atoms or molecules approaching the tip are ionized in its vicinity, either on the way in or on the rebound. The ions are accelerated along the lines of force toward the

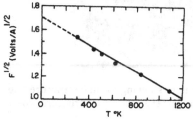

Fig. 52. Graph of $F^{\frac{1}{2}}$ versus temperature for the field desorption of thorium from tungsten at low coverage in a constant time of 3 sec. Data from Müller.[3]

negative screen so that a greatly magnified image of the ionization zone is produced there.

Since the probability of ionization is a sensitive function of field, the greatest current will come from regions where this is highest. The field distribution near the tip is determined by its atomic structure. To a first approximation the equipotentials can be approximated by sections of spheres, centering on the surface atoms, as shown in Fig. 53. At x_c, where ionization occurs, the overlap of the equipotential spheres causes a certain amount of smearing out and loss of structure, compared to the tip. Field anisotropies will therefore be most marked for atoms with relatively few nearest neighbors and will be least for atoms densely packed in planes. Consequently the former may show up on the pattern as discrete points of high "emission" while the latter may not show up at all at the threshold field. As the voltage is raised, some of the latter regions will begin to emit. At even higher fields ionization becomes appreciable everywhere and the pattern will gradually lose all structure. Optimum resolution will therefore be obtained at fields slightly in excess of the threshold for the most exposed regions of the tip surface.

Resolution

The resolution depends on two factors, first, the transverse velocity v_t of image-forming ions and second, the loss of detail in the field distribution at x_c. As pointed out by Müller,[2a] it is possible to minimize v_t by total or partial thermal accommodation on the tip before ionization, so that the incoming thermal and polarization velocity compo-

Fig. 53. Schematic cross section of a field emitter showing atomic structure and its effect on the equipotentials near the tip. The latter show most structure where atoms are least closely packed.

nents are reduced. Under these conditions Eq. (2.19) yields in the usual way

$$\delta = 4r\beta(kT/Ve)^{\frac{1}{2}}. \tag{77}$$

For $r = 10^{-5}$ cm, $\beta = 1.5$, and $V = 10^4$ v, Eq. (77) shows a resolution of 10 A at 300° K and 2.4 A at 20° K.

The second factor is more difficult to assess quantitatively, since a fine-grained knowledge of the charge distribution on the tip would be required. The problem may be handled in a way leading to the correct functional dependence and a reasonable numerical estimate by stipulating that two atoms will just be resolved if the equipotential spheres of radius x_c drawn about them intersect only "slightly." For the condition of tangency the resolution would be $2x_c$, so that the "slight" intersection can be handled by stipulating that

$$\delta = 2\epsilon x_c, \tag{78}$$

where ϵ is a numerical constant of the order of 0.5 to 1.0.

Under conditions where transverse velocities are small, that is, at low temperature, the ultimate resolution is given by Eq. (78) and seems to be $\leqslant 4$ A. It is interesting to express the dependence on F and I implied by x_c more explicitly. To a reasonable approximation, ionization depends on the value of D given by Eq. (3), that is, it will occur when

$$F \cong gI^{\frac{3}{2}}, \tag{79}$$

where, from experiment, $g = 3.8 \times 10^6$ for F in volts per centimeter and I in volts. Substitution in Eq. (1) shows that

$$x_c \cong \frac{I - \phi}{3.8 \times 10^6 I^{\frac{3}{2}}}, \tag{80}$$

so that the resolution is

$$\delta \cong 53\epsilon \frac{I - \phi}{I^{\frac{3}{2}}} \text{ A.} \tag{81}$$

This brings out the dependence of δ on the ionization potential of the image-forming gas.

The optimum resolution will therefore depend on a delicate balance of the following factors: (1) small transverse and high forward velocity; (2) small x_c; (3) high contrast, that is, optimum anisotropy in ionization probability on "high" and "low" spots. Factors (1) and (2) imply minimum temperature consistent with desorption, and maximum accelerating voltage, hence field. Factor (3), on the other hand, implies values of field near the mid-point of the curve shown in Fig. 45. This requirement opposes the first two to some extent.

The balance seems to be most favorable for helium at 20° K. The lower resolution obtainable with hydrogen may in part be due to a secondary reaction leading to H^+ with some of the excess energy appearing transverse to the flight direction.

Figure 54 shows helium-ion images of a tungsten and a platinum tip, obtained by E. W. Müller. The resolution is so good that a major fraction of the surface atoms can be resolved. This is probably the most direct proof of the discreteness and regularity of an atomic lattice ever obtained and is a truly remarkable achievement.

Fig. 54(*a*). Helium-ion micrograph of 011 oriented tungsten field emitter.
(Courtesy of Professor E. W. Müller.)

APPLICATIONS OF FIELD IONIZATION AND FIELD DESORPTION

Field-Ion Microscopy

Perhaps the most striking application of field ionization is the ion microscope,[2a] whose theory and method of operation have already been discussed. To date it has been used principally to examine with very high resolution the lattice structure of clean field-emitter surfaces. Thus Müller has been able to show that tips cleaned by heating still have considerable surface roughness in the form of superficial

Fig. 54(b). Helium-ion micrograph of 100 oriented platinum field emitter. (Courtesy of Professor E. W. Müller.)

atoms on top of the completed lattice planes. Field emission provides only hints of this from the fact that the work function of the 110 plane of tungsten seems to depend on the heat treatment of the tip.[11]

When the field is increased to the point where field desorption of the lattice itself occurs at room temperature, perfect surfaces free from thermal disorder can be obtained.[2a] The ion micrograms shown in Fig. 54 are of this kind.

Since the resolution of He and even H_2 ion micrograms is good enough to resolve lattice planes, it is possible to obtain fairly accurate tip radii by counting the number of planes between two crystallographic directions. Field desorption of the lattice itself always occurs from the edges of planes, where the field is highest, and can thus be observed in ion microscopy as the shrinking and collapsing of the ring-shaped steps.

Probably the most interesting application of ion microscopy will be the study of imperfections and overgrowths. Thus Müller[20] has been able to see individual lattice vacancies on planes where the atomic packing is loose enough to permit the resolution of individual atoms. He could even form an estimate of the volume density of vacancies by field-desorbing the lattice, layer by layer, until a new vacancy appeared.

A large number of similar applications present themselves, although these have not as yet been realized. Thus it may be possible to examine a clean tip for screw dislocations and other imperfections, oxidize or corrode it *in situ*, and look for correlations between the lattice imperfections and the most active reaction sites. (See Appendix 2.)

The limiting factor of field-ion microscopy is that fields of 2 v/A are required for hydrogen-ion and 3.8 v/A for helium-ion work so that the emitter surface is subjected to surface stresses of the order of 2×10^{10} to 10^{11} dy/cm². At the highest stresses field evaporation of the lattice itself occurs [2,3] for the lower-melting metals even if the bulk strength is adequate, so that one is frequently limited to hydrogen-ion work. As with ordinary field emission, the bulk strength of emitters could be greatly improved by using vapor-grown whiskers.

Mass-Spectrometric Applications

The use of a positively charged field emitter as ion source in a mass spectrometer[1] offers a number of fascinating applications, most of which have not yet been exploited very fully. Some of these are discussed below.

Adsorption. It has been pointed out that field desorption can be distinguished from field ionization by the use of pulsed fields and by the variation of peak shape with field. Thus the dissociation product in the case of CD_3OH adsorbed on the 110 plane of a tungsten emitter was shown to be CD_3O^{+12}. This technique should permit many detailed studies of dissociative adsorption on single-crystal planes, since the spectrometer normally sees only a very small fraction of a given face of the emitter. In the case of very simple adsorption mechanisms, it is possible to determine individual rate constants from the variation of products with pulse-repetition rate.

Even if the kinetics are too complicated for this analysis, much useful information can be obtained. Thus the degree of coverage relative to its equilibrium value at any given pressure and temperature can be found as a function of pulse-repetition rate. If desorption at zero field is negligible, this yields information very similar to the so-called flash-filament technique,[13] but over a more precisely determined region of the substrate. Various modifications of the method are evidently possible.

Since field desorption occurs most readily for moderately adsorbed species, the method should find its most fruitful application in the study of catalytic reactions. Fairly weak binding is precisely the condition required for desorption of products and continuation of the catalytic processes.

Interesting applications are also possible in the case of physical adsorption. Thus water polymers were found at pressures of 10^{-3} mm-of-mercury and $300°$ K when direct-current fields were used.[1] Since H_2O has a large permanent dipole moment, the field-dipole interaction increases the binding energy sufficiently to cause multilayer adsorption. Since this is rather liquid-like, a high degree of polymerization is to be expected.

Secondary Processes. Table 2 lists various secondary ions from hydrocarbons, occurring at relatively low abundance. In most cases their intensity depends on the square of the gas pressure and they show up with fractional masses corresponding to the geometric mean of the parent and final species. This indicates that they were formed by collision after final acceleration. Since the ions pick up 90 percent of their final energy in a distance of the order of $10r_t$, this is precisely what one would expect for low pressures and long paths.

This suggests some interesting possibilities for studying the reactions of fast ions with various gases. It is easily possible to adjust ion energies after formation so that reaction cross sections can be studied as functions of energy. In the case of CH_4, for instance, the collisional removal of one to four H atoms seems about equally probable at ~ 5000 v.

Occasionally secondaries (characterized by their fractional mass) whose intensity depends on the first power of the pressure are found at very low abundance. These must be due to breakup without collision.

Film Reactions. Recent experiments by Beckey [8] indicate that liquid-film formation, discussed on p. 81, leads to a number of very unusual effects. Thus the water polymers reported by Inghram and Gomer [1] seem in fact to be $H_3O^+ \cdot (H_2O)_n$ because of the reaction

$$H_2O + H_2O \rightarrow H_3O^+ + OH^- \tag{82}$$

at the layer surface. Similarly he finds, under conditions where a methanol film exists, that the reaction

$$2CH_3OH \rightarrow CH_3OH_2^+ + CH_3O^- \tag{83}$$

seems to occur. In the case of H_2O layers, the reactions

$$H_2O + M^+ \rightarrow MH^+ + OH \tag{84}$$

also seem to occur whenever the ion M^+ can complete an electron shell by hydride formation.

The foregoing indicates that a number of very interesting and unusual chemical reactions can be studied by this technique.

Analytical Applications. Field ionization leads to virtually no vibrational or electronic excitation, so that the ions are quite stable. Consequently spectra do not show the complexity associated with electron impact, since there is very little breakup. It is further possible to distinguish primary ions from those formed either by collisional or vibrational excitation on the one hand or by dissociative adsorption on the other. An example of the resulting simplicity is the field-ion spectrum of acetone, shown in Fig. 55. There is only one major peak (apart from isotopic ones), in contrast to the electron-impact spec-

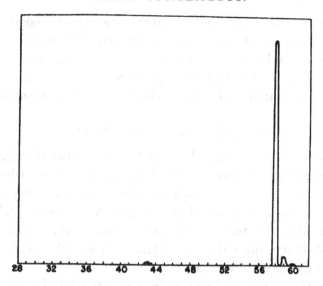

Fig. 55. Field-ion spectrum of acetone.

trum, which boasts 18. The appearance fields depend sensitively on ionization potentials, as shown by Eq. (3), so that the analysis of isomeric mixtures is not impossible.

In addition to the more or less conventional analytical applications indicated above, field-ion sources can be used as detectors with unusual properties. Thus small concentrations of free radicals produced in thermal or photochemical reactions can be picked up since there is no interference from the parent substance. The lifetime of excited states may be estimated by the following scheme. The voltage is so adjusted that field ionization occurs from the excited but not from the ground state. The excitation source (for example, a light beam) is placed at varying distances from the tip. If the velocity of particles is known, the decay times can then be found.

The various mass-spectrometric applications of field ionization and desorption suffer from two drawbacks. (1) The currents ordinarily obtainable are of the order of 10^{-12}–10^{-17} amp. This necessitates the use of electron multipliers in conjunction with sensitive electrometers as detectors. It is possible to devise electron-optical systems for collimating the ion beam and obtaining higher intensities. This is not suitable for desorption studies if the effect of crystal structure is important. (2) Changes in tip radius during an experiment are often encountered.

The reason for this may be secondary negative-ion bombardment of the tip region or field-induced reactions with the substrate.

Only the second of these possibilities can be regarded as serious, but means of circumventing it can very likely be found. Thus Schissel [7] has noted that graphite whiskers make very stable tips for ion work.

Other Applications of Field Desorption

Tip Cleaning. It was pointed out in Chapter 1 that the cleaning of conventional field emitters is limited by the heating to which they can be subjected. Müller has used a combination of heating and field desorption to clean emitters of such metals as platinum and iron.[14] Field desorption also tends to remove "loose" surface atoms of the emitter substance, since these experience higher fields, and leads to highly idealized crystal surfaces.

Investigation of Potential Curves in Adsorption.[4] It has been mentioned that the field desorption of thorium from tungsten seems to obey kinetics consistent with an image-potential adsorption curve. Analogous measurements enable one, in principle, to determine the shape of the ground-state curves in covalent binding. The magnitude of Q in Eq. (60) depends on the shape of $V(x_c)$. One would therefore determine Q as a function of F by measuring rates of desorption as functions of temperature at a number of field strengths. The values of $V(x_c)$ so found can then be correlated with the corresponding x_c values by solving Eq. (72) for x_c. (See Appendix 2.)

CHAPTER 4

Some Applications of Field Emission to Adsorption

The previous chapters have dealt with the theory of field and ion emission and have described the principles of the corresponding microscopes. Only a brief sketch of actual and potential applications was given. The value of this book will be enhanced if some of these are discussed in more detail, since this will enable the reader to judge for himself the strengths and weaknesses of the technique. This chapter and the concluding one are therefore devoted to selected specific applications, although no attempt at inclusiveness has been made.

In the author's opinion, some of the most fruitful applications of field emission lie in the study of adsorption. Before describing some of these, a short discussion of the nature of physical and chemical adsorption will be presented.

GAS-SOLID ADSORPTION

There are two general types of adsorption on solids. One of these, physical adsorption, is very nonspecific and occurs on all solids with all gases. The forces are of the van der Waals type, of fairly long range, but only weakly attractive. Adsorption energies are generally only slightly higher than the corresponding heats of liquefaction, adsorption is appreciable only in a temperature-pressure range near condensation, and many layers may be usually formed reversibly.

The other type, chemisorption, is much more specific. Energies of adsorption vary from one to many volts, and this tight binding is confined to one or at most part of a second atomic layer. This may be stable even at high temperature and zero pressure, so that chemisorption can be highly irreversible. The magnitude of the forces indicates electronic rearrangement. Often this can be accomplished only at the cost of severe modification or outright dissociation of the adsorbed molecules or by chemical reaction with the substrate, leading to permanent changes in the latter. Langmuir's [1] classic work indicates that hydrogen or oxygen is adsorbed as atoms on most metals and can be desorbed in this form at sufficiently high temperatures.

Chemisorption corresponds to chemical binding and the act of adsorption to a chemical reaction. Since the electrons or holes contributed by the solid are very mobile, at least in metals, and since any surface cut from a discrete periodic array of atoms presents a variety of atomic configurations, chemisorption is less specific than ordinary molecular binding. This manifests itself not only in a wide variety of adsorbate-adsorbent pairs, but also in variations of the adsorption energy with coverage and orientation in a given system.

After this brief introduction, the two types will be discussed separately, although it is clear that some overlap is possible.

Physical Adsorption

The existence of multilayer adsorption was first postulated by Brunauer, Emmett, and Teller,[2] who noted that many isotherms showed inflection points at pressures of about $0.1P_0$, P_0 being the condensation pressure. Beyond this point a slow rise occurred in the amount adsorbed, followed near P_0 by a very steep increase (Fig. 56).

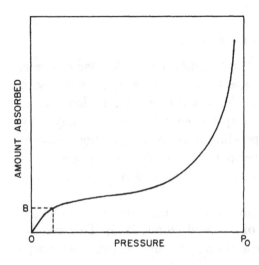

Fig. 56. BET isotherm, showing inflection point B.

They explained the initial inflection (point B) as corresponding to the completion of a monolayer; this is followed by multilayer adsorption and finally, near P_0, by condensation of liquid. They incorporated these ideas in the BET theory of adsorption, which leads to a mathematically simple isotherm. However, as was first pointed out by Halsey,[3] the model on which it is based is a highly idealized one. Basically it is equivalent to assuming simultaneous partial formation of many layers. Since it is physically unreasonable to expect the stacking of isolated molecules on top of each other, multilayer formation on a homogeneous surface should proceed in steps, a given layer having a finite population only after the near completion of the previous one. Stepwise adsorption of this kind has in fact been observed for inert gases on tungsten field emitters.[4] On most adsorbents there is not only a wide range of actual surfaces exposed but also a great deal of pore structure and capillarity at points of contact between particles. This leads to a smearing out of the idealized step isotherms and results in the quasi-BET behavior of many real systems.

Although the theoretical inadequacies of the BET model are apparent, its identification of the inflection point B with the completion of the first layer is often qualitatively correct. This is the case because the heat of adsorption in this layer is always somewhat larger than in subsequent ones, often sufficiently so to break through the masking caused by the heterogeneity of the adsorbent. In any case, point B can legitimately be identified with a change in adsorption energy. This fact lends enormous practical utility to the BET isotherm as a means of at least estimating the real surface area of adsorbents, catalysts, and substrates for chemisorption, since a size which is only slightly arbitrary can be assigned to the adsorbate.

The correct formulation of the statistical-mechanical equations of physical adsorption is not difficult but the solution has so far proved impossible without drastic simplifications. The liquid-slab model of Halsey, Hill, and Frenkel [5] aims at accomplishing this by treating physisorption as the formation of a liquid-like slab on the surface. Isotherms of the type

$$\ln (P/P_0) = -k/V^r \tag{1}$$

result and seem to hold experimentally near P_0 with $r \cong 3$, V being the volume adsorbed and k a constant.

The major unsolved theoretical problem in physisorption is one of statistical mechanics. The magnitude and nature of the forces is fairly well understood in many, though not all, cases. On the experimental side there is a great need for studying homogeneous, or at least completely characterized, substrates in order to avoid the smearing out already mentioned. Recent work [4,6] using field emitters as adsorbents seems to indicate at least one promising line of attack and has produced some food for theoretical thought.

Among other things, these studies seem to show that multilayers may remain liquid-like considerably below the bulk melting point of the adsorbate.[4] Probably this is connected with the inhibition by the substrate lattice of long-range order in the adsorbate.

Before leaving the subject, one recently emerged but incompletely understood fact will be mentioned. It was first noted by Mignolet,[7] and subsequently confirmed by field-emission work,[4,6] that physical adsorption of inert gases and of molecules on metals produces a large dipole moment (0.1–0.8 debye) in the adsorbate, the negative end pointing toward the surface. Despite this fact, adsorption energies appear to be "normal." It is interesting to speculate on the mechanism of the dipole formation (charge transfer, polarization?) and on the reasons for the lack of severe anomalies in the resultant heats.

Chemisorption on Metals

From the work of Langmuir,[1] Roberts,[8] Beeck,[9] and more recent investigations, the following picture of chemisorption on metals is emerging, although there are many important unanswered questions.

Adsorption is rapid and occurs with almost zero activation energy in most cases,[1,8–12] despite the common occurrence of dissociation. Sticking coefficients are high, of the order of 0.1–0.5, and vary with coverage.[11,12] It seems probable that they differ from unity because re-evaporation of physically preadsorbed molecules can precede chemisorptive binding.[12] Since ad-sites seem to be definite crystallographically determined locations, impinging molecules must diffuse to these before becoming chemisorbed, so that re-evaporation will occur more frequently at higher coverages where a longer search is required. This leads to a variation in sticking coefficient with coverage and temperature. A knowledge of the diffusion coefficients and activation and evaporation energies of physically adsorbed molecules on clean metals is therefore important also to an understanding of the kinetics of chemisorption.

Above a certain point there may also be a real scarcity of adjacent pair sites, as was first pointed out by Roberts,[8] so that some molecular chemisorption may occur in the case of normally dissociating molecules. Field-emission studies indicate that approximately 80 percent of the chemisorbed layer can be formed with no activation energy at 4–20° K for hydrogen or oxygen on tungsten or nickel. In the case of hydrogen there seems to be a reversible equilibrium, at very high coverage and below 4.2° K, between the molecular and a portion of the atomic adsorbate.[10] True activation energies of dissociative adsorption on clean metals seem to occur only with carbon monoxide, for which it is so high that it does not occur at all, as first found by Ehrlich,[13] and with nitrogen, for which this author found evidence (with the field-emission microscope) for physical adsorption at 77° and chemisorption at 300° K. The reason for this is most probably that the intersection of the potential curves for pure chemisorption and pure physisorption occurs at a point where the latter is still and the former already attractive, so that splitting of the levels results in a smoothly attractive ground-state curve (Fig. 57).

Relatively little reliable work had been done on the kinetics of desorption. The processes are obviously activated, the energy corre-

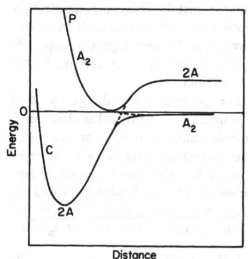

Fig. 57. Schematic cross section through potential-energy surfaces for physisorption P and chemisorption C, showing the separation leading to nonactivated dissociative chemisorption.

sponding to the thermodynamic heats of adsorption in those cases where there is no activation energy of *ad*sorption. It is known from mobility studies (using field emission) that surface mobility almost always precedes desorption, so that measured activation energies represent averages over the surface. These are heavily weighted in favor of the sites of tightest binding participating in the process.

Generally, atomic desorption (in the case of polyatomic end products) requires much higher energies than molecular desorption and therefore occurs only at much higher temperatures, despite the fact that it requires the correct synchronization of the vibrations of the participating atoms. Thus H_2 evaporates rapidly from tungsten at 600–700° K,[10] while atomic desorption is inappreciable below 1000° K.

One of the striking characteristics of most chemisorption systems is the fact that the heat of adsorption decreases with coverage.[9] Studies of surface diffusion by means of the field-emission microscope [10] seem to explain this behavior, at least in part.

It is found that diffusion occurs with activation energies ranging from $\sim$ 10 to $>$ 45 percent of the adsorption energies and that both this ratio and the absolute magnitudes depend on the atomic surface configuration. The values increase in both cases with surface roughness (on the atomic scale). It is also found that sites of inherently different binding energy occur on one and the same crystal face (except perhaps the most closely packed) as an unavoidable consequence of the discrete atomic structure of the metal. This suggests strongly that decreases in heats of adsorption result from this built-in heterogeneity, at least at moderate coverages. At very high coverage, adadparticle interactions can become important and may account for the atomic-molecular switching already noted.

Contact-potential measurements (among other things by field emission) indicate that binding is usually covalent rather than ionic, although moderately large bond dipole moments are common. Attempts to correlate these (or the contact potentials directly) with heats of binding have been made,[14] but must be taken with a grain of salt because (*a*) many of the old values are in considerable error and (*b*) there is no one heat of adsorption in a given system. The criterion for covalent or ionic binding on metals is the sign of the inequality

$$H_{at} \gtrless \phi - I + H_{ion} + 3.6q^2/x, \qquad (2)$$

where H_{at} and H_{ion} are the "intrinsic" adsorption energies of atom and ion in volts, ϕ and I the work function and ionization potential, and the last term on the right an image-attraction term depending on the effective ionic charge and the ion-surface separation x (in angstroms). This term is unlikely to be accurate at very small separations, but no calculation of its quantum-mechanical analogue is presently available. On metals the small effective dielectric constant results in polarized but covalent adsorption for most substances of high vacuum ionization potential. As already pointed out in Chapter 3, field desorption is often able to distinguish experimentally between ionic and covalent binding.

While the direction of electron transfer in ad-bonding is usually indicated by the sign of the contact potential (when correctly obtained), other methods have also been used. Changes in paramagnetism[15] or electric conductivity[16] have been interpreted along these lines. Unfortunately, it is necessary to use evaporated films for these methods. These are very microcrystalline and sometimes amorphous, so that there is a serious question of the validity of the interpretations even if the experimental conditions are adequate. It would be interesting to measure conductivity changes on adsorption in a monocrystalline metal whisker of very small volume-to-surface ratio.

The detailed nature of the ad-bond in chemisorption is very poorly understood. There have been relatively few and usually highly simplified quantum-mechanical calculations.[17] In view of the lack of detailed knowledge of clean metal surfaces, this is hardly surprising. While we are quite unable to make a priori energy or dipole-moment calculations (even on the semiempirical level of molecular spectroscopy), certain qualitative statements can be made.[18] Strong adsorption will occur only if substrate and adsorbate respectively have available electrons and orbitals. Transition metals fit particularly well into the roles of both donor and acceptor because of their partially filled d-bands, since these are narrow and correspond to fairly localized electrons. The d-electrons have a tendency to lower their energies by bonding, which enables them to approach the bonding nucleus with a consequent decrease in potential energy. This is true only of d-electrons in partially filled bands. As filling nears completion, there is more overlap and exchange within the band and hence less need for outside alliances.

One may regard ad-bonds as localized impurity states somewhere below the Fermi level. Demotion of an electron to this state is energetically most profitable if it came from the vicinity of the Fermi level. This is possible for d-electrons in partly filled bands, but may be endoenergetic in filled ones. The s-electrons form broad nonlocalized bands and therefore have relatively little to gain from localization by bonding.

This picture seems to explain the observed behavior [9] fairly well, but one must add a word of caution. Many experimental heats of adsorption, particularly on low-melting metals, are quite suspect because of cleanliness and vacuum conditions. Further, it is usually difficult to measure negative molecular heats of adsorption even though these may still correspond to quite high atomic adsorption energies.

The effect of surface structure, so intuitively obvious in terms of discrete bonds or neighbor interactions, can be justified quantum-mechanically on the basis of the uncertainty principle. Minimum localization and consequent lowering of the kinetic energy of the ad-particle electron result from the largest number of participating substrate atoms.

Adsorption on Semiconductors

This chapter is largely concerned with emphasizing the role of field emission in the elucidation of surface phenomena. Since very little work of this nature has actually been performed on semiconductors, only the briefest outline of adsorption on them will be given.

As with metals, chemisorption occurs and involves electron transfer. Ionic binding can occur, at least at low coverage, since the large dielectric constant of most semiconductors lowers the effective ionization potential or electron affinity of the adsorbate. If the electrons involved come from or go to the interior of the substrate (either its conduction band or impurity levels), a space-charge region will be set up. This Schottky layer has already been mentioned in Chapter 1, p. 10. Such a layer creates a barrier for electron flow *to* electron-acceptor or *from* electron-donor adsorbates, and results in an activation energy for adsorption. In addition, the strong layer potential set up by ionic adsorbates leads to a rapid decrease in the heat of adsorption of further ions by an amount

$$-\Delta H = 4\pi l q^2 N_s \theta, \tag{3}$$

where q is the ionic charge, l the layer spacing, N_s the number of ad-sites per unit area, and θ their fractional coverage. Ionic adsorption will therefore stop when the heat of adsorption at $\theta = 0$, H_0 equals this decrease, or when

$$\theta = \frac{H_0}{4\pi l q^2 N_s}. \tag{4}$$

Since this value of θ is quite low, covalent binding is likely to occur also, particularly if a discrete set of levels or states exists at the surface with which the adsorbate can interact.

The presence of physically adsorbed molecules or atoms capable of accepting or donating electrons constitutes per se a type of surface state, as pointed out in Chapter 1, p. 24. In addition to this, the clean semiconductor surface should also possess a set of intrinsic levels, differing from those of the bulk. These Tamm states are due to the finite size of the specimen and can be thought of as arising from the "impedance matching" at the surface, or, more physically, as the unsatisfied bonds of the surface atoms.

Unfortunately, the presence of even small amounts of impurities may give rise to simulated intrinsic surface states, so that the experimental situation is not very clear.

It may be possible to elucidate some of the points just raised by field emission. Thus the transition from the screened to the field-penetration region in emission depends on the number of surface states. It may be possible to investigate this as a function of adsorption, and in particular to see whether there is *any* screening with a really clean surface.

It would also be very interesting to obtain information on the relation between surface structure and adsorption, for instance by mobility studies.

Practical Considerations

The specificity and differentiation encountered in chemisorption make the use of clean and if possible monocrystalline substrates an absolute necessity if the results of an investigation are to have any meaning. Once a clean surface has been obtained, its continued purity depends on the pressure in the system. If a sticking coefficient of 0.1 is assumed, monolayer formation takes 1 sec at 10^{-6} mm-of-mercury, so that pressures of 10^{-9} mm-of-mercury or lower are necessary. It is not particularly difficult to achieve this [19] if the system can be baked out at 430° C and is free from volatile impurities. However, once gas is admitted to a conventional system, the reattainment of high vacuum requires either very long times or another bake out.

A rather gentler procedure,[10] free from this objection, consists in immersing a sealed-off system in liquid helium or hydrogen. Since the vapor pressure of all gases except helium and possibly hydrogen is $\leqslant 10^{-15}$ mm-of-mercury at 4.2° K, ultrahigh vacuum is attained merely by cooling. It is also possible to include relatively high gas pressures (STP) in the system so that exposure or re-exposure of a substrate can be accomplished by warming up. High vacuum is then attained by recooling. The sticking coefficients of all gases are very high at low temperature. This fact can be utilized for diffusion studies, as indicated below.

The attainment of clean surfaces, apart from vacuum requirements, has been discussed briefly in Chapter 2. As pointed out, heat cleaning is often sufficient. For low-melting metals it is necessary to grow field emitters under high-vacuum conditions in the form of whiskers. This will be discussed in the next chapter.

After this brief account of adsorption phenomena and the role of field emission in their elucidation, a somewhat more detailed description of some of these experiments follows.

SURFACE DIFFUSION

The preceding section has indicated the information obtainable from diffusion studies. Some of these will be described and discussed here.

The principle of the method is extremely simple and consists in evaporating the gas to be investigated onto a field emitter in such a way that only a part of the latter receives a deposit.[10] Since all adsorbates change the emission, it is then possible to heat the tip and to follow the ensuing surface diffusion processes visually or with a motion-picture camera.

Unilateral gas deposition is accomplished by taking advantage of the high sticking coefficients of all gases at very low temperature. A sealed-off field-emission tube is immersed in liquid helium or hydrogen, and the tip is cleaned electrically and allowed to cool. Gas is then evaporated from an electrically heatable source. Since rebounds from the walls are negligible, and the ratio of tip to wall area is very small, only those portions of the field emitter in the direct line of sight of the gas source receive a deposit. By varying the source-tip geometry the deposit region can be varied, as shown in Fig. 58.

Gas sources either utilize the thermal decomposition of such compounds as CuO or ZrH_2 (formed *in situ* [10]) or depend on the evaporation of gas selectively precondensed on an electrically heatable platinum foil.[4,20] In the latter case the emission tube contains 10–30 mm-of-mercury of the gas to be studied. This can be condensed over and over again. Figure 59 shows a schematic diagram of a tube for diffusion studies and Fig. 60 a condensation gas source.

Accurate temperature measurement and control can be achieved with a servomechanism (Fig. 61) whose sensing element compares the voltage in the heating loop with that produced in a resistor in series with it. The latter is set to the desired value of R, obtained from a previous R-versus-T calibration. The servomechanism automatically and rapidly adjusts the heating current until the resistances of loop and control are equal. In most experiments, radiation from the tip is small or negligible so that its temperature may be taken as that of the apex of the support loop.

Diffusion and Adsorption of Hydrogen and Oxygen on Tungsten [10a,c]

The results for these gases are qualitatively similar, although there are significant differences. The activation energies and temperatures

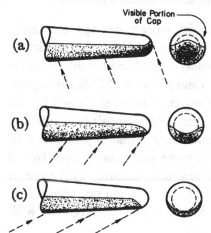

Fig. 58. Schematic diagram of deposit geometries on a field emitter: (*a*) source ahead of tip; (*b*) source aft of tip; (*c*) the beam misses the visible portion (broken circle in head-on views) entirely.

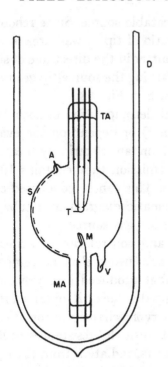

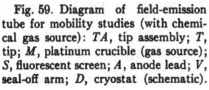

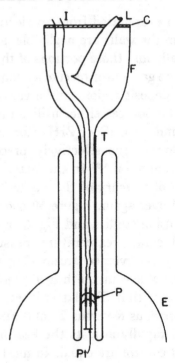

Fig. 59. Diagram of field-emission tube for mobility studies (with chemical gas source): *TA*, tip assembly; *T*, tip; *M*, platinum crucible (gas source); *S*, fluorescent screen; *A*, anode lead; *V*, seal-off arm; *D*, cryostat (schematic).

Fig. 60. Schematic diagram of universal gas source for diffusion studies: Pt, electrically heatable platinum sleeve; W, tungsten leads; *P*, press seal; *D*, Dewar seals; *T*, Teflon sleeve; *F*, storage funnel for liquid helium or hydrogen; *L*, electric leads to source; He, helium or hydrogen inlet tube.

at which a given phenomenon occurs with oxygen are approximately double those for hydrogen. At least three types of diffusion occur.

(1) For deposits in excess of a monolayer, a sharp boundary, moving almost uniformly over the tip, is observed at ∼ 27° K for oxygen (Figs. 62–70) and at ≪ 20° K for hydrogen. The layer formed in this way (Figs. 71 and 72) is not mobile; if the initial deposit is insufficient for complete spreading, the sharp boundary remains stationary to very much higher temperatures unless more gas is deposited. In the latter case, movement is resumed at low temperature. This type of diffusion also shows an upper limit above which it will not occur. This is ∼ 70° K for oxygen and too low for accurate measurement with hydrogen.

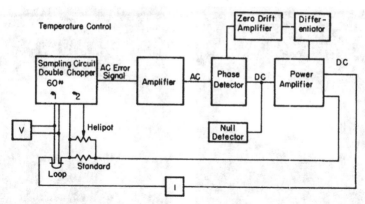

Fig. 61. Schematic block diagram of tip-temperature controller: I, precision ammeter; V, precision millivoltmeter. *Loop* indicates the tip-support loop; the voltage fed into the sampling circuit is that developed between the potential leads (see Fig. 59).

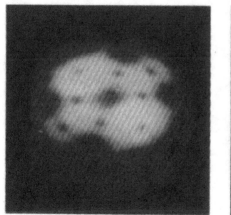

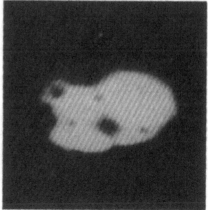

Fig. 62. Pattern from clean tungsten tip.

Fig. 63. Same tip, with oxygen deposit, $\theta > 1$, at 4.2° K.

These facts strongly indicate the following mechanism. At low temperature the chemisorbed layer is immobile. However, gas physically adsorbed on top of this layer is mobile and can wander to the edge of the chemisorbed deposit, become trapped there, and so extend the monolayer. This results in further migration over the newly covered region and leads to an extension of the chemisorbate. Since there is a sharp discontinuity in coverage at the monolayer-clean tungsten edge, a sharp boundary is observed. Boundaries of this kind

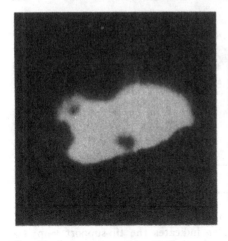

Fig. 64. Same tip, after 0.43 sec at 27° K.

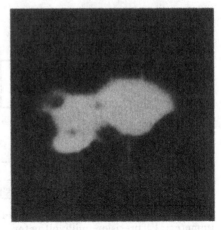

Fig. 65. Same tip, after 1.28 sec at 27° K.

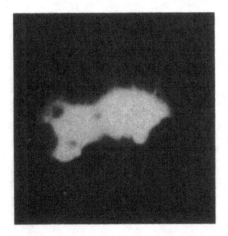

Fig. 66. Same tip, after 2.28 sec at 27° K.

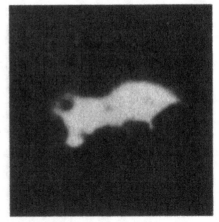

Fig. 67. Same tip, after 3.57 sec at 27° K.

invariably result from diffusion with precipitation if the mean precipitation or trapping distance is less than the resolution of the system. In this case the precipitation distance is of the order of a lattice spacing, namely, $\sim 3A$, while the resolution of the field-emission microscope is 20–30 A.

The observed upper temperature limit corresponds to the evaporation of physically adsorbed molecules before their migration to the edge of the monolayer and chemisorption on the bare surface can occur.

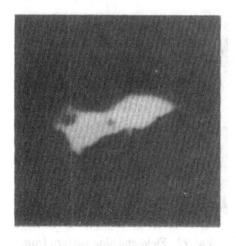

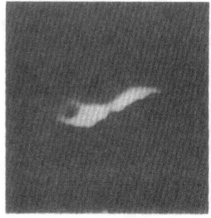

Fig. 68. Same tip, after 5.28 sec at 27° K.

Fig. 69. Same tip, after 9.30 sec at 27° K.

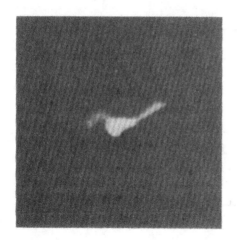

Fig. 70. Same tip, after 12.6 sec at 27° K. The bright region corresponds to the oxygen-free remnant of tip. Figures 62–70 are taken from a 16-mm motion picture of the diffusion process.

The distance traversed by the boundary in time t is approximately

$$x = (Dt)^{\frac{1}{2}} \qquad (5)$$

(Fig. 73). If the diffusion coefficient D is assumed to have the form

$$D = a^2\nu \exp\left(-E_d/kT\right), \qquad (6)$$

ν being a frequency of the order of 10^{12} sec^{-1} and a a jumplength (~ 3 A), E_d can be estimated. In the case of oxygen this turns out to be 0.9 kcal.

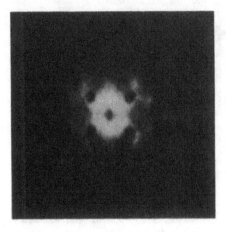

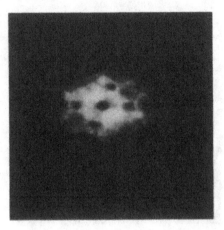

Fig. 71. Field-emission pattern from a tungsten tip fully covered with hydrogen; $\phi = 5.03$ v.

Fig. 72. Field-emission pattern from a tungsten tip covered with oxygen; $\phi = 6.2$ v.

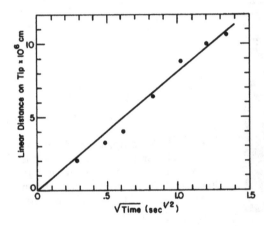

Fig. 73. Graph of average distance traversed by boundary in type 1 diffusion of oxygen on tungsten at $27°$ K as a function of $t^{\frac{1}{2}}$; tip radius, 2.10^{-5} cm.

The mean life τ of diffusing oxygen molecules is approximately

$$\tau = \nu^{-1} \exp (E_{ad}/kT) \text{ sec.} \tag{7}$$

If the frequency appearing in this expression is assumed to be equal to that for surface diffusion, Eqs. (5), (6), and (7) may be combined to give

$$E_{ad} = E_d + 9.2T \log_{10} (\bar{x}/a), \tag{8}$$

where $\bar{x}$ is the mean distance traversed by a diffusing molecule before evaporation. Equation (8) permits an estimate of the adsorption energy if E_d and $\bar{x}$ are known. The latter can be determined from the field-emission patterns as a function of T if identical initial deposits are used. In this way $E_{ad} = 2.8$ kcal was estimated for oxygen.

Qualitatively similar results have been found for CO [20] and CO_2.[21] The temperature and energy ranges also correspond roughly to values found in the physical adsorption of inert gases of corresponding molecular weight.[4,6]

It is interesting that the coverage after type 1 diffusion is of the order of 80 percent of the saturation value, as shown by the average work-function changes. This indicates that at most 20 percent of the chemisorbed layer requires appreciable activation energies for its formation. It is possible to show that some rearrangement in the initial low-temperature deposit occurs with oxygen at $\geqslant 300°$ K. If another dose of oxygen is then deposited on the cooled tip, type 1 diffusion occurs again and results in a higher work function and coverage than after the first spread. This indicates that some of the sites nonbonding at 30° K become filled by local rearrangement at 300° K thus liberating more bonding sites to be filled on the next low-temperature spreading.

(2a) For initial deposits of 0.8 to 1 monolayer a moving boundary diffusion is observed between 180 and 240° K with hydrogen (Figs. 74–80) and 500–530° K with oxygen (Figs. 81–84). The activation energy of these processes has been established fairly accurately by Arrhenius plots as 5.9 kcal for hydrogen and 25.5 kcal for oxygen.

In contrast to type 1 diffusion, the boundary observed here does not move uniformly over the tip but spreads radially outward from

74   75   76

Fig. 74. Clean tungsten tip with hydrogen deposit ($\theta = 0.8$) in lower half.

Fig. 75. Same tip after heating to 190° K.

Fig. 76. Same tip after further heating to 190° K. The boundary almost reaches the 112 and 121 faces (see Fig. 16b for indexing).

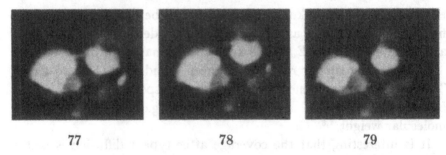

77 78 79

Fig. 77. Same tip after further heating to 190° K. The boundary
has reached 112 and 121.

Fig. 78. Same tip after further spreading at 190° K. The boundary starts to
broaden laterally in the channels connecting 011 with 112 and 121.

Fig. 79. Same tip after further heating to 190° K. The boundary has now reached
101 and 110 via the channels 011–112–101 and 011–121–110. The face 211 (upper
right-hand corner) is just being engulfed from 101.

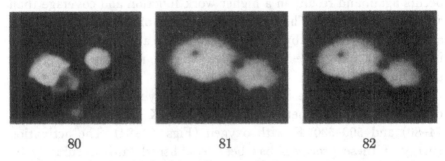

80 81 82

Fig. 80. Same tip after further spreading at 190° K. The 111 face is now isolated
(upper bright spot) and so is 010 (bright region on left-hand side).

Fig. 81. Tungsten tip with oxygen deposit ($\theta = 0.8$) in lower half.

Fig. 82. Same tip after 5 sec at 517° K. The boundary is beginning to spread
from the 011 face and has reached 112.

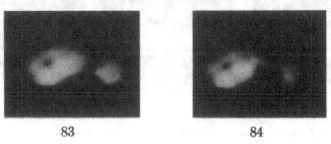

83   84

Fig. 83. Same tip after 20 sec at 517° K. The boundary is reaching the 121 face.

Fig. 84. Same tip after 70 sec at 570° K. The deposit moves laterally in the
channel 011–121.

the central 110 face (Figs. 74–84). It can be seen that it advances most rapidly along zones like 110–211.

A precipitation mechanism can again be postulated to account for the sharp boundary. However, both the temperatures and the energies make it virtually certain that processes within the chemisorbed layer are involved. Figure 85 shows that the 110 face is the closest packed in a bcc lattice, but is surrounded by atomically rough

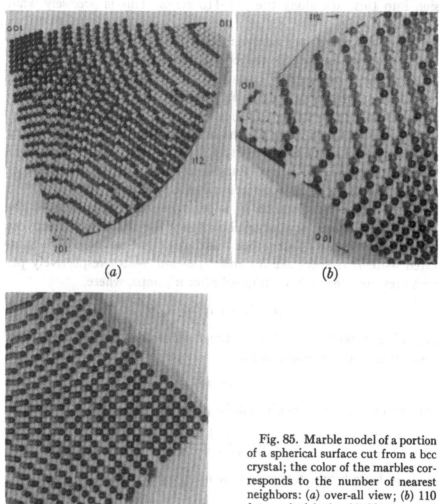

(a)

(b)

(c)

Fig. 85. Marble model of a portion of a spherical surface cut from a bcc crystal; the color of the marbles corresponds to the number of nearest neighbors: (a) over-all view; (b) 110 face and vicinals; (c) 001 face and vicinals.

surface in all directions except zones like 110–211, and so forth. The latter consist entirely of terraces of 110 orientation. It is reasonable to assume not only least binding but also maximum mobility on the most tightly packed face of a given structure. The central 110 face and the special zones linking it to other 110 faces constitute a network of low-impedance paths for diffusing atoms. If the initial deposit should not include the central 110 face one would expect to see diffusion into this face along the 211–110 zones. This is precisely what happens.

Atoms will therefore migrate rapidly to and over the central 110 face of the tip, but will be trapped on the rough surface at its edge. The trap sites presumably consist of surface configurations offering more near neighbors to a diffusing atom than the 110 face. However, local saturation of traps (t-sites) still leaves many short low-impedance paths open, since all regions of the surface consist to varying degree of small portions of 110 faces, stacked in different orientations with respect to each other. Further migration over these regions (d-sites) can proceed if excess adsorbate is available.

If the trapping distance x_t is less than the resolution of the microscope, a sharp boundary will be seen at the discontinuity between clean and artificially "smoothed" surfaces (that is, those with saturated t-sites). If there are N_d and N_t d- and t-sites respectively per unit area, an atom will be trapped after n jumps, where

$$n = (N_d + N_t)/\gamma N_t \tag{9}$$

and γ is a trapping coefficient, of the order of unity. If a random-walk assumption can be made, so that

$$x/a = n^{\frac{1}{2}}, \tag{10}$$

the precipitation distance x_t will be given by

$$x_t = a\left(\frac{N_t + N_d}{\gamma N_t}\right)^{\frac{1}{2}}. \tag{11}$$

A sharp boundary will therefore be observed if

$$\frac{N_t}{N_d + N_t} \geqslant \frac{1}{\gamma}\,(a/\delta)^2, \tag{12}$$

where δ is the resolution.

If diffusion from d- to d-sites rather than from trap to trap is to be rate-controlling, the coverage θ_f after spreading must be

$$\theta_f \geqslant \frac{N_t}{N_t + N_d}. \tag{13}$$

In this way it can be estimated that ~ 40 percent of all sites on atomically rough regions of the tip act as traps in the case of hydrogen and ~ 60 percent in the case of oxygen. Equation (11) shows x_t to be of the order of 3–4 A, consistent with the requirements of Eq. (12) for the observance of a boundary and with the physical interpretation of trap sites.

($2b$) In the case of oxygen another type of boundary spreading is also observed. For initial deposits of $\theta \geqslant 0.4$, a boundary spreads outward from the cube faces at $\sim 400°$ K with an activation energy of 22 kcal (Figs. 86–88). After the cessation of this process an increase in the temperature to $500°$ K produces "ordinary" type 2 diffusion with the boundary moving radially outward from the 110 face for sufficiently high deposits (Fig. 89). This suggests that a certain fraction of the adsorbed O atoms are less tightly bound on 100 than on 110. The explanation probably lies in the large size of oxygen, which may enable a single O atom to fill the interstice between four W atoms in a 100 plane (Fig. 85c), thereby creating a surface on which migrating O atoms can make contact with only two W atoms at a time, thus being less tightly bound than on 110 where contact with three W atoms is always possible.

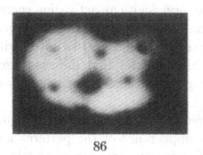 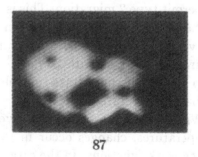

86   87

Fig. 86. Field-emission pattern from a tungsten tip with oxygen deposit ($\theta = 0.8$) in the lower quarter.

Fig. 87. Same tip after heating to $407°$ K for 8 sec. Note the boundary spreading outward from 001.

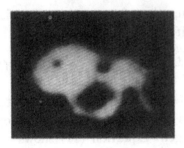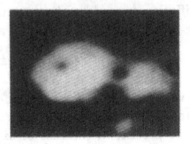

Fig. 88. Same tip after heating to 518° K for 10 sec. The boundary from 001 has reached the 112 face.

Fig. 89. Same tip after heating to 544° K for 30 sec. Diffusion into and radially outward from 011 has occurred and stopped because all mobile O atoms have been precipitated.

One may ask why hydrogen does not show similar behavior. Again the answer seems to be connected with size. Hydrogen atoms are so small that all regions of the surface seem either 110 or trap-like. Different planes should therefore differ to a first approximation only in the number but not in the type of trap sites and should show the same activation energy for type 2 diffusion. This is the case. The boundary moves only half as rapidly along 110–111 as along 110–100, but there is no noticeable difference in E_d. The surface is atomically much rougher near 111 than near 100.

(3) The preceding arguments imply boundary-free diffusion, corresponding to migration from t- to t-site, for deposits insufficient to permit type 2 migration. This is observed with an activation energy of 9.5 kcal in the case of hydrogen and 30 kcal in the case of oxygen. At very low coverages a further rise to 16 ± 3 kcal is noted with hydrogen, but not with oxygen. This indicates that diffusion from trap to trap becomes rate-controlling when there is not enough adsorbate to saturate the traps.

When hydrogen- or oxygen-covered emitters are heated to higher temperatures, changes occur in the appearance of the patterns and in the work functions. In the case of hydrogen these can be associated with desorption; the changes can always be reversed by further H_2 deposition and diffusion; the desorptive stages correspond in appearance and work function to those produced by the equilibration of various initial H_2 deposits. It is possible to measure the rates as a function of temperature and to obtain a series of activation energies of desorption for various average coverages. Since the activation energy of adsorption is very low, if not zero, that for desorption can be equated with the heat of adsorption with respect to the stable gas-phase species. Desorption occurs rapidly at 600–700° K, so that the latter is certainly H_2 rather than H.

The diffusion results indicate mobility at the desorption temperatures. The measured heats of desorption are therefore averages over the surface. However, the tightest-binding sites that must be emptied in a given coverage interval will weight this heavily. This can be seen by considering desorption as diffusion out of a trap site to a d-site and evaporation (with a partner) from the latter. The total activation energy (or ΔH) is the sum of the energies for both processes.

The results obtained in this way are in reasonably good agreement with calorimetric data,[9] except for the presence of a high-energy tail of 60 kcal at very low coverage (Fig. 90). It has been pointed out by Hickmott [22] that so high a value is incomptatible with the observed desorption rates at 700° K. At present there is no very good explanation for this discrepancy. The value of 60 kcal may refer to some phenomenon other than desorption, conceivably a slight local re-arrangement of some W atoms.

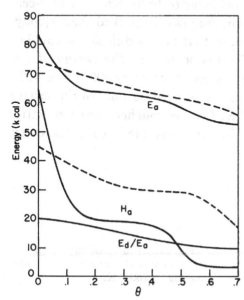

Fig. 90. Heat of adsorption of H_2 (H_a) and energy of binding of H atoms on tungsten as function of coverage θ. Solid curves obtained by field emission; [10] broken curves by calorimetry. [9] The lowest curve gives the ratio of activation energy for diffusion to binding energy; for this curve the ordinate reads in percent.

In the case of oxygen, the changes that occur on heating above 600–700° K are irreversible in the sense that redeposition of oxygen does not lead to the original appearance or work functions. It must therefore be concluded that desorptive changes are accompanied by reaction with the substrate. This leads to the formation of definitely oriented oxide layers with local field enhancement and consequent high emission (Fig. 91). It is necessary to distinguish between increases in emission due to this cause and those due to desorption (with work-function decreases). Müller [23] has been able to show with the field-ion microscope that the bright regions in patterns like Fig. 91 are due to field enhancement rather than low work function, thus confirming the redeposition results.

The adsorption of oxygen on W and the pattern changes on heating have also been investigated by J. A. Becker,[24] who interprets them in terms of desorption. Flash-filament experiments indicate that there is *some* desorption on heating, although recent results indicate that the desorbent may be in part CO. However, the interpretation of what is *left* on the tip depends on the fate of the oxygen remaining behind.

Diffusion and Adsorption of Hydrogen on Nickel [10b]

It is interesting to study the behavior of hydrogen on a face-centered substrate, since this is much more closely packed than a body-centered one. Results on nickel show that type 1 diffusion occurs but cannot be observed directly. The reason is that the energy of impinging H_2 molecules from the hot gas source ($\sim 300°$ K) is sufficient to permit them to skate over the covered portions of the surface even when the latter is at 4.2° K. On the rougher tungsten lattice this phenomenon can also occur since it is possible to cover the entire

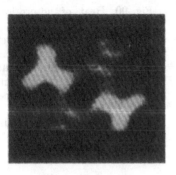

Fig. 91. Field-emission pattern from oxygen-covered tungsten tip after heating to 1300° K.

tip with hydrogen or oxygen without heating, but only by using much larger doses.

Boundary diffusion in the chemisorbed layer does not occur on nickel except vestigially near the 110 faces. The activation energy of the boundary-free diffusion is 7 ± 1 kcal, which is almost the same as that for type 2 diffusion on tungsten. The reason for this behavior is most probably the close packing of the nickel lattice. Figure 92

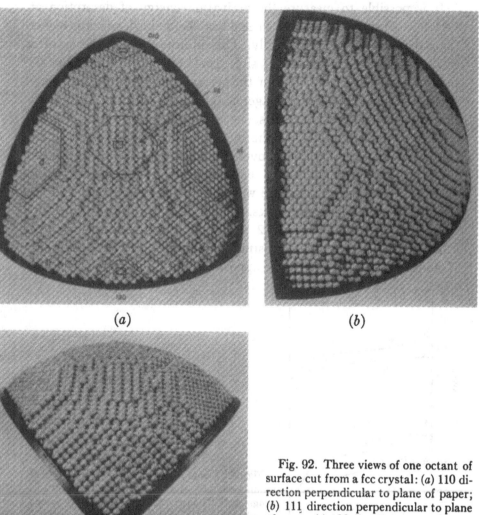

(a)

(b)

(c)

Fig. 92. Three views of one octant of surface cut from a fcc crystal: (a) 110 direction perpendicular to plane of paper; (b) 111 direction perpendicular to plane of paper; (c) 100 direction perpendicular to plane of paper.

shows that any region on the surface consists of steps and terraces of close-packed material, either 111 or 100, except in the immediate vicinity of the 110 faces, where slightly looser configurations exist. The packing of atoms even there is tighter than on the most close-packed face of tungsten. Thus there are virtually no trap sites to be filled, so that low-energy diffusion can occur without precipitation everywhere except very near 110. Application of Eq. (12) with $\delta = 30$ A and $a = 3$ A shows that $N_t/N \leqslant 0.01$.

It is possible to measure the activation energy of desorption at very low coverage. A value of 46 ± 4 kcal is obtained, which is higher than that found calorimetrically. This indicates the presence of a small number of tight binding sites, probably near 110. Figure 93 shows a curve of H_a and E_{ad} for the system. It is seen that these are flat over a large range of coverage, in agreement with the conclusions drawn from the diffusion results.

The relative uniformity of the nickel surface is also indicated by the appearance of hydrogen-covered nickel tips. Unlike tungsten, there is very little additional emission anisotropy on adsorption (Fig. 94), although the average work-function increments (~ 0.5 v) are almost identical in both cases.

At very low temperatures (2–4° K) and maximum coverage, an interesting phenomenon suggesting equilibrium between adsorbed H and adsorbed H_2 occurs. It appears that the heat of adsorption

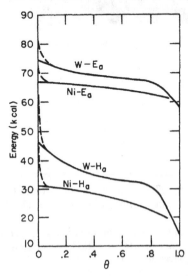

Fig. 93. Heats of adsorption of H_2 (H_a) and binding energies of H atoms (E_a) on nickel and tungsten.[9] Broken portions represent values found at very low coverage by field emission.[10]

Fig. 94. Field-emission pattern from 111 oriented nickel tip: (*left*) clean, (*right*) hydrogen-covered; $\phi = 5.0$ v.

with respect to H_2 is so small at this coverage that field-dipole interaction energies can shift the $(2H \leftrightarrow H_2)_{ads}$ equilibrium. The details of the argument [10b] are involved and will not be given here.

Adsorption of Carbon Monoxide on Tungsten [13,20,25]

This case is particularly interesting for two reasons. It was found by Ehrlich [13] that adsorption occurs without dissociation, since no traces of carbon are found after desorption. Thus one is dealing with molecular chemisorption. Second, the large size of the adsorbate provides a useful variation in that parameter. Only preliminary information is available at present. All three types of diffusion occur. Migration in the chemisorbed layer starts below 700° K and is rapid over the 110 face and the zones connecting it with 211. The activation energy of this process is $\sim$ 21 kcal. Sharp boundary diffusion occurs and tends to surround the 100 faces and their vicinals, with the boundary closing in on the former. The activation energy of this process is $\sim$ 36 kcal. After the cessation of boundary diffusion type 3 migration occurs ($E_{diff} \cong 70$ kcal), between 900 and 1000° K but is accompanied by desorption. The latter is rapid at 1300° K.

Examination of a marble model of the emitter shows that the tightest-binding sites for CO occur on 100 configurations, a site corresponding to the entire interstice between four corner atoms of a unit cell face. It is evident that diffusion over a filled row of such sites is impossible. Thus the boundary is first noted when the adsorbate reaches portions of the surface where these configurations begin to be numerous (that is, on the approaches to 100). As the latter is neared, trap sites predominate to the point where there are no diffusion paths between them. At this point the activation energy corresponds to diffusion out of trap sites. Since this involves the rupture of almost half the ad-bonds, the activation energy is so high that desorption also occurs.

It is significant that boundary diffusion occurs here even though there is no dissociation. This rules out molecular diffusion with desorptive dissociation as the mechanism of type 2 diffusion. This is a possible although highly unlikely event in the case of hydrogen or oxygen. In this connection it should be added that the results with inert gases on clean tungsten rule out the high heats of adsorption of undissociated H_2 or O_2 on clean tungsten required by Eq. (8) for this hypothetical mechanism.

Adsorption of Carbon Dioxide on Tungsten [21]

This system is interesting chiefly because it illustrates the potentialities of field emission for the study of simple heterogeneous reactions. When carbon dioxide is evaporated onto one side of a tungsten emitter, diffusion sets in at 400° K. The appearance and activation energy suggest that this is the 100 oxygen migration. At $\sim$ 500° K, type 2 110 oxygen migration is observed, followed at $\sim$ 600° K by type 3 migration. The activation energies and pattern changes are those corresponding to oxygen diffusion. At 800° K a diffusion resembling that of carbon monoxide occurs. On further heating, oxidation of the tip results, leading to the usual oxide patterns.

These findings suggest that CO_2 is dissociatively adsorbed as CO and O. Since the diffusion temperatures for these substances differ, their presence on the substrate can be established. It would be interesting to know whether CO_2 is dissociated below 400° K where the first migration occurs. This question could be answered by a mass-spectrometric field-desorption experiment.

Adsorption and Diffusion of Inert Gases on Tungsten [4,6]

The methods outlined at the beginning of this section make it possible to evaporate arbitrary doses of inert gases unilaterally onto field emitters and to study their subsequent behavior on heating.[4] The results are qualitatively very similar for all gases studied. With initial deposits of a monolayer or less, boundary-free diffusion occurs on heating with concomitant enhancement of emission. This is most pronounced in the regions surrounding 100 (Fig. 95). Its activation energy can be estimated from Equations (5) and (6).

When larger amounts are evaporated, the region of deposit is darkened. Under these conditions boundary free diffusion occurs at the same temperatures as before and again leads to enhanced emission on the originally clean portions of the tip. Further rises in temperature now produce diffusion with sharp boundaries. These occur in definite waves or steps, and it is possible to see two or more of these simultaneously inundating the tip (Fig. 96). Each succeeding wave leads to a decrease in emission, and to loss of resolution (Fig. 97).

Further heating leads to evaporation. This occurs with sharp, almost concentric boundaries, shrinking toward the tip's center at different rates. The innermost boundary, enclosing the darkest region, shrinks most rapidly, and the outermost, corresponding to highest emission, least so (Fig. 98). After the disappearance of the last boundary, the first layer, corresponding to enhanced emission, remains. Its desorption occurs only at slightly higher temperatures, and, except in the case of neon, without a boundary. It is possible to estimate the activation energy of the process from the temperature dependence of the rates and Eq. (7).

The qualitative meaning of these observations is clear. At sufficiently low temperatures, (multilayer) physical adsorption occurs.

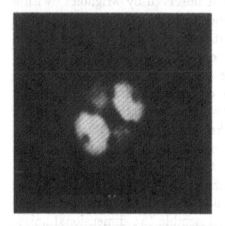

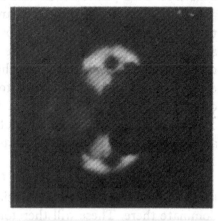

Fig. 95. Field-emission pattern from tungsten tip covered with monolayer of argon. Note enhanced emission, especially around 100 faces; $\phi = 3.7$ v.

Fig. 96. Layers of argon flowing over tungsten tip at $\sim 20°$ K. Note boundaries.

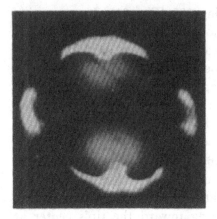

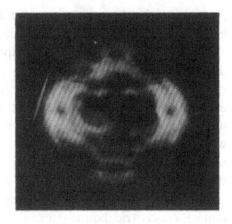

Fig. 97. Field-emission pattern from tungsten tip covered with multilayers of argon. Note lack of structure in the pattern, corresponding to decreased resolution. Bright regions at the edges of the pattern correspond to lower argon coverage. The deposit is piled up on the tip by the tidal pull of the interaction between field and dielectric.

Fig. 98. Multilayers of neon evaporating from a tungsten field emitter at $\sim 5°$ K. Note the concentric rings corresponding to various rates at which the layers shrink.

The formation of the first layer is accompanied by a decrease in work function, probably because of a charge-transfer complex with the substrate. This phenomenon was first observed by Mignolet [7] with a zero-field method and therefore cannot be associated with a polarization due to the external field (apart from having the wrong sign in any case). The magnitude of the contact potentials indicates large dipole moments per ad-atom in the first layer. The resultant dipole-dipole repulsions are sufficient to overcome attractive forces and cause this layer to behave like a two-dimensional gas. The absence of these repulsions in neon, where the contact potential is very small, allows attractive forces to dominate so that even the first layer shows cohesion.

Since the mechanism of dipole formation is of short range, higher layers do not participate appreciably so that cohesive forces predominate there. These will therefore resemble two-dimensional anisotropic liquids, whose boundaries result from line tension, the two-dimensional analog of surface tension. Evaporation should occur from the layer edges where an atom is least tightly held. Since the distance of a layer from the substrate determines its binding energy, the highest one should evaporate most rapidly. It will be shown later that emission decreases with the thickness of the adsorbed film. The behavior illustrated by Fig. 98 can therefore be explained as the simultaneous evaporation of different layers at varying rates, the highest layer evaporating most rapidly, and the first one least so.

In all cases the liquid-like flowing of boundaries is observed well below the bulk melting point of the adsorbate. This can probably be explained as follows. The substrate lattice structure is transmitted to the adsorbate layer by layer. Up to a point this keeps the latter from assuming its own bulk structure over large distances and thus prevents long-range order. Accordingly the layers behave like liquids, but to different degrees. Beyond a certain distance from the surface the disordering effect of the substrate begins to cost more in the loss of ad-adparticle interaction than is gained by ad-substrate attraction. The multilayered, liquid adsorbate consequently becomes metastable with respect to crystallization. The latter is observed indirectly at low temperature where spontaneous breaks in the low-emission adsorbate occur. The formation of patches of high emission suggests the invisible piling up of adsorbate as crystallites. Slight heating melts these and heals the breaks in the liquid.

The adsorption of inert gases on tungsten and tantalum field emitters has also been studied by Ehrlich,[6] who admitted the gases at low pressure to a field-emission tube, keeping only the tip cold. With the exception of neon, which was not investigated by him, and those results obtainable only by unilateral gas deposition, his experimental results are in complete agreement with those described above.[4] Ehrlich prefers to interpret the monolayer contact potentials as arising from the polarization of adsorbed atoms by the field existing at the metal surface, in agreement with Mignolet's[7] original hypothesis. The author feels that the extent of the surface dipole layer, 0.5–1 A, is too small to "seize" more than a fraction of each ad-atom, whose distance from the surface corresponds to a van der Waals dimension, that is, several angstroms. However, no unequivocal decision is possible at present. Ehrlich has carried out a detailed analysis of the nearest-neighbor fit of inert-gas atoms with various regions of the surface and finds remarkably close agreement between the regions of maximum-emission enhancement and those of best fit. His experiments enable him to conclude that the maximum-emission enhancement is largely due to increased adsorbate concentration in the vicinity of 100, rather than to an increased dipole moment per ad-atom. The details of the reasoning are given in reference 6.

A number of interesting conclusions can be drawn from the work-function changes connected with multilayer adsorption. These will be discussed in Chapter 5. At this point all the results on surface diffusion for chemically and physically adsorbed gases will be summarized.

Summary of Surface-Diffusion Results

Table 4 shows the results obtained so far. Diffusion coefficients and apparent activation energies E_d (app.) are based on Eqs. (5) and (6). Where separate values of E_d are available from Arrhenius plots, the comparison with E_d (app.) permits the estimation of an activation entropy, ΔS^*, based on the assumption of a jump length of 3 A and a frequency of 10^{12} sec^{-1}.

The most important results of Table 4 are the relation of E_d to

TABLE 4. Summary of surface-diffusion results.[a]

Type of diffusion	$a^2 \nu e^{\Delta S^*/R}$ (cm^2/sec)	E_d (kcal)	ΔS^* (eu)	E_{des} (kcal)	E_d/E_{des}
CO on W boundary free	—	65 ± 5	—	90 ± 5	0.72
CO on W boundary	—	[36]	—	[80]	[.45]
O on W boundary free	82	30 ± 1.5	13 ± 5	130	.24
O on W 110 boundary	3×10^{-2}	24.8 ± 1	7 ± 5	125	.2
O on W 100 boundary	1	22.7 ± 1	13 ± 5	125	.18
H on W boundary free	3.2×10^{-4}	$9.6\text{–}16 \pm 3$	$[-2 \pm 5]$	65–82	.20
H on W 110 boundary	1.8×10^{-5}	5.9 ± 1	$[-8 \pm 5]$	60	.1
H on Ni boundary free	3.2×10^{-5}	7 ± 1	-7 ± 5	68–72	.1
CO_2 on CO_2/W	$[10^{-3}]$	2.4	—	5.5	.43
CO on CO/W	—	[0.9]	—	[2.3]	.39
O_2 on O/W	$[10^{-3}]$	.9	—	2.3	.39
Xe on W	$[10^{-3}]$	[3]	—	9–10	.3
Kr on W	$[10^{-3}]$	[0.9]	—	5.9	[.18]
A on W	$[10^{-3}]$	.6	—	1.9	.3

[a] Column 2 gives the preexponential part of the diffusion coefficient. Column 4 lists activation entropies. The symbols X/W refer to an X-covered W surface. Quantities in brackets refer to preliminary values or rough estimates.

E_{ad} and the variation of this ratio with the nature of the adsorbate and substrate. It will be noted that E_d/E_{ad} increases with E_d, that is, with surface roughness. This is entirely reasonable. The potential structure of the surface imitates to some degree its physical one. On a perfectly smooth substrate diffusion would require zero activation energy, regardless of the energy of adsorption. As the roughness increases, so does the potential corrugation. Consequently, place change will require more and more partial desorption.

A similar trend is noticeable in going from hydrogen to carbon monoxide. This is probably connected with the adsorbate size and the number of formal bonds made with the substrate. The small size of an H atom permits it to be in contact with almost its full normal complement of W atoms even during diffusion. With O, and even more with CO, which may be attached to the surface with both atoms, diffusion requires more partial desorption.

These findings indicate that E_d/E_{ad} cannot be predicted quantitatively from pairwise nearest-neighbor interactions, except perhaps with very large or physically adsorbed particles, although such considerations are very useful in predicting and corroborating qualitative behavior.

The ratio of E_d/E_{ad} for physisorbed gases falls midway in the chemisorption range. Since van der Waals forces are active in this case, a model based on pairwise interactions with nearest and next-nearest neighbors should be particularly good.[4,6] It is therefore interesting that the results fall in the same range as those with chemisorbed gases where such a model is more suspect.

It must be concluded that a certain degree of heterogeneity is built into all but the most closely packed faces of any crystal. Since the microstructure consists of various combinations of a very small number of different building units, the same types of sites can appear on many different faces, the variation being often chiefly in number. Thus macroorientation can be much less important than one might at first suppose. The extent to which a given adsorbate notices the surface heterogeneity depends on its size, since this determines both the number and the position of substrate atoms an ad-particle can interact with simultaneously. Thus the effects of structure would disappear for infinitely small adsorbates since it is always possible to pick three surface atoms whose relative position is that found on 110 (for instance). Hydrogen on tungsten begins to approach this situation, although the variation in the electronic configuration of the surface on 110 compared with 111 or 100 is sufficient to bind H more tightly on the latter two and to increase E_d/E_{ad} there. However H, unlike O or CO, is too small to differentiate very much between the configurations occurring on 100 and 111.

While these results are far from complete, they indicate the type of information obtainable by field emission.

CHAPTER 5

Some Miscellaneous Applications of Field Emission

This chapter is intended to show the range and variety of field emission by treating a number of separate topics, ranging from the quantum mechanics of emission through dielectric layers to the kinetics of whisker growth and including the still puzzling subject of molecular images.

FIELD EMISSION THROUGH DIELECTRIC LAYERS

The emission of electrons from tungsten covered with inert gases was described in the last chapter from the point of view of adsorption. The phenomenon is also interesting from the emission standpoint and this aspect will be considered here.[1]

Table 5 summarizes the apparent work function and emission changes resulting from the adsorption of inert gases on tungsten field

TABLE 5. Summary of emission data for inert gases on tungsten.[a]

Gas	ϕ_{mono} (ev)	$\log a_W/a_{mono}$	apparent ϕ_{multi} (ev)	$\log a_W/a_{multi}$
Neon	4.35	0.8	4.05	2.2
Argon	3.70	1.1	4.30	1.5
Krypton	3.35	1.5	3.40	2.0
Xenon	3.20	0.2	3.00	0.5

[a] The a's refer to the preexponential part of the Fowler-Nordheim equation as written in Eq. (2.38). The subscripts on all quantities refer to the condition of the emitting tip: W, clean tip; mono, tip covered with one layer of gas; multi, tip covered with a thick deposit.

emitters. For heavy coverages the apparent work function decreases for neon, increases for argon, and stays more or less constant for krypton and xenon, relative to the monolayer values. In all cases the total emission is decreased because of changes in the preexponential term.

Mechanism

The thickness of the adsorbed film is 20–30 A at maximum coverage and thus exceeds the length of the potential barrier seen by electrons near the Fermi level, even if allowance is made for the reduction in field by $1/K$, where K is the dielectric constant listed in Table 6.

TABLE 6. Physical constants and pertinent data for inert gases

Gas	Density [a] (gm/cm^3)	Dielectric constant, K	Polarizability, α_p (10^{-24} cm^3)	Crystal radius (A)
Neon	1.2$_{27}$	1.23	0.392	1.59
Argon	1.65$_{40}$	1.55	1.65	1.91
Krypton	2.1$_{120}$	1.56	2.50	2.01
Xenon	2.7$_{133}$	1.82	4.10	2.20

[a] Subscripts denote absolute temperatures to which densities refer.

Since electrons from the vicinity of the Fermi level contribute most heavily to field emission, the barrier region of interest can be considered to be filled with inert gas atoms. Only the first layer of these contains strong dipoles, so that emission occurs through a shell of essentially nonpolar dielectric. It will therefore be assumed that it occurs by tunneling through a potential barrier corresponding to monolayer conditions, but modified as follows: (1) the field within the dielectric shell is reduced by $1/K$; it can be shown from elementary electrostatics that this will be the case if its thickness is small compared with the tip radius; (2) the image potential is reduced by $1/K$; (3) the potential of electrons in the dielectric is reduced by an average polarization energy; (4) each inert-gas atom represents a short-range but deep attractive potential for electrons, because of the incomplete screening of the nucleus at small distances; the barrier opacity is reduced by the presence of these potential holes.

The first two effects will tend to raise the apparent work function, while the last two will reduce it. The qualitative correctness of this picture is indicated by Table 5, which reflects the competition of dielectric and hole effects. In the case of neon, which has a low dielectric constant, the hole effect predominates, in argon the opposite is the case, and in krypton and xenon the effects more or less balance each other.

The situation can therefore be treated as follows. The apparent work function must be divided by $K^{\frac{2}{3}}$ to correct for the locally decreased field. The image-potential correction can be handled by multiplying this value by $(\alpha_w/\alpha_{multi})^{\frac{2}{3}}$ in accordance with Eq. (2.41). The derivation of the image correction term shows that α_{multi} can be found from Table 1, as a function of the modified argument

$$y = 3.8 \times 10^{-4} F_{vac}^{\frac{1}{2}}/K\phi, \qquad (1)$$

where F_{vac} is the field that would exist if there were no dielectric. The appropriate ϕ and α can therefore be found by iteration, along the lines described on p. 48.

The polarization energy $\Delta\phi_{pol}$ can be readily calculated [1b] as a lattice sum over the terms $\frac{1}{2}\alpha_p F_{eff}^2$, where α_p is the atomic polarizability and F_{eff} the effective field at an atom due to the polarizing electron. The result is a small correction to the energy except at very close atom-electron distances. There the effect is going to be lumped in with the hole strength of the atoms. The work-function change due to the hole effect can now be found by subtracting the corrected multilayer work function from that of the monolayer:

$$\Delta\phi_{hole} = \phi_{mono} - \phi_{multi} - \Delta\phi_{pol}. \qquad (2)$$

The results are shown in Table 7.

This treatment is valid only if the monolayer contact potential is unaffected by subsequent adsorption. This is almost certainly the

TABLE 7. Summary of hole parameters for inert gases.[a]

Gas	$\Delta\phi_{hole}$ (ev)	a (A)	r_0 (A)
Neon	0.80	0.46	0.60
Argon	0.50	0.42	0.52
Krypton	0.89	1.23	1.00
Xenon	1.17	1.95	1.38

[a] $\Delta\phi_{hole}$ is the work-function change due to the hole effect; a is the scattering length; r_0 is the hole radius.

case, as shown by the close agreement of its field-emission and zero-field values, which excludes the importance of hole effects for a single layer. This is not surprising, since plausible mechanisms for the production of large dipoles in that layer exist. It should also be pointed out that emission *through* atoms (which will be important in multi-layer adsorption) will occur only when decreases in field make this the path of lowest impedance despite scattering losses. This is probably not the case for the monolayer.

The real problem has now been isolated. It consists in finding the effect of an array of small potential wells on the expectation value of the work function.

This can be done in two approximations. The first, utilizing a method invented by Fermi,[2] treats the situation as a scattering problem and converts the experimental work function decreases into scattering lengths. The other treats the problem as one-dimensional and comes up with physical dimensions for the potential wells.[1] If a given well shape is assumed, theoretical values for the Fermi scattering lengths can be computed from the well dimensions and compared with the experimental scattering lengths found by the first method.

Since this represents a rather unusual and interesting quantum-mechanical application of field emission, the arguments will be presented in some detail.

Fermi's Treatment of Holes in a Slowly Varying Potential

The following ingenious method was developed by Fermi in 1934 to account for the term shifts in alkali-vapor spectra in the presence of inert gases.[2] If small holes of depth V and radius r_0 are distributed with uniform density in a slowly varying potential U, the wave function ψ found from the solution of

$$\nabla^2\psi + (2m/\hbar^2)(W - U)\psi - (2m/\hbar^2)\sum_i V_i\psi = 0 \qquad (3)$$

will vary slowly except near a hole. Let us define an average wave function $\langle\psi\rangle$. This will resemble ψ closely except near a hole, where it will be free from its violent variations. Since

$$\nabla^2\langle\psi\rangle = \langle\nabla^2\psi\rangle, \qquad (4)$$

$$\nabla^2\langle\psi\rangle + (2m/\hbar^2)(W - U)\langle\psi\rangle - (2m/\hbar^2)\left\langle\sum_i V_i\psi\right\rangle = 0. \qquad (5)$$

This is a completely general expression. The problem now consists in finding a way of evaluating the last term in Eq. (5). Near a hole ψ depends mainly on V, since U varies slowly with distance. We may therefore choose the center of a given hole as origin and set up the wave equation in terms of the distance r from this point. If the hole is spherically symmetric and if only low-energy, that is, s-states are considered, the equation can be simplified in the customary manner by introducing a function $u(r)$,

$$u(r) = r\psi, \tag{6}$$

so that

$$u''(r) + (2m/\hbar^2)(W - U - V)u(r) = 0. \tag{7}$$

Since the holes are deep and W small, this becomes

$$u''(r) = (2m/\hbar^2)V(r)u(r) \qquad \text{for } r \leqslant r_0 \tag{8}$$

and

$$u''(r) = 0 \qquad \text{for } r > r_0, \tag{9}$$

if the zero of energy is chosen at the edge of the hole. The solution for $r > r_0$ is

$$u(r) = kr + c, \tag{10}$$

so that

$$\psi = k + c/r. \tag{11}$$

As r increases, ψ approaches the limit k since $c/r \rightarrow 0$. However, far from a hole ψ approaches $\langle\psi\rangle$. Therefore

$$\langle\psi\rangle \cong k \tag{12}$$

and

$$u(r) = r\langle\psi\rangle + c \qquad \text{for } r > r_0, \tag{13}$$

or

$$u(r) = (r + a)\langle\psi\rangle, \tag{14}$$

where a is a constant with the dimension of length. Figure 99 illustrates the situation. The function $u(r)$ outside the hole is linear in r and extrapolates to an intercept a on the abscissa. This quantity is called the Fermi scattering length. Inside the hole $u(r)$ behaves as shown and vanishes at $r = 0$ to at least the order of r, since ψ must remain finite. At $r = r_0$, $u(r)$ matches the linear portion smoothly as required. This fact will be utilized later in obtaining an expression for a in terms of the properties of V. Had the origin $r = 0$ been chosen

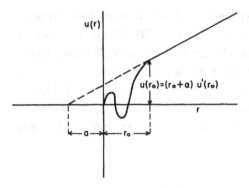

Fig. 99. Diagram showing the behavior of the function $u(r)$ versus r near a potential hole of radius r_0; a, Fermi length.

outside a hole, $u(r) = r\langle\psi\rangle$ would have gone to zero at $r = 0$. The presence of the hole therefore has the effect of shifting this zero by an amount a.

We are now able to evaluate the summation term in Eq. (5). This is

$$
\begin{aligned}
(2m/\hbar^2)\left\langle \sum V_i\psi \right\rangle &= (2m/\hbar^2)n\int_0^\infty V\psi \, d\tau \\
&= \frac{2mn}{\hbar^2}\int_0^{r_0} 4\pi r^2 \frac{Vu(r)}{r} \, dr + 0 \\
&= 4\pi n\int_0^{r_0}(2m/\hbar^2)Vu(r)r \, dr,
\end{aligned}
\tag{15}
$$

where n is the density of holes. On making use of Eq. (8) this becomes

$$
(2m/\hbar^2)\left\langle \sum_i V_i\psi \right\rangle = 4\pi n\int_0^{r_0} u''(r)r \, dr.
\tag{16}
$$

Integrating by parts we have

$$
(2m/\hbar^2)\left\langle \sum_i V_i\psi \right\rangle = 4\pi n[u'r - u]_0^{r_0}.
\tag{17}
$$

At the lower limit both r and $u(r)$ vanish, so that we obtain, with the help of Eq. (14),

$$
(2m/\hbar^2)\left\langle \sum_i V_i\psi \right\rangle = -4\pi na\langle\psi\rangle.
\tag{18}
$$

This can be substituted in Eq. (5) to yield finally

$$
\nabla^2\langle\psi\rangle + (2m/\hbar^2)\left(W - U + \frac{2\pi nah^2}{m}\right)\langle\psi\rangle = 0.
\tag{19}
$$

It is seen that this has the form of an ordinary wave equation if the total energy W is replaced by an energy W' given by

$$
W' = W + 2\pi nah^2/m.
\tag{20}
$$

The solutions of Eq. (19) will have the exact form of the unperturbed ones except that all energies must be modified as indicated. The validity of the treatment hinges, once again, on the assumptions of low energy and small holes.

In the present case a formal solution in terms of $\langle \psi \rangle$ would result in the Fowler-Nordheim equation with a Fermi level raised by $2\pi nah^2/m$. This apparent increase in electron energy can just as well be regarded as a decrease in work function, so that we finally have

$$| \Delta\phi_{\text{hole}} | = 2\pi nah^2/m. \tag{21}$$

The scattering lengths found from Eq. (21) are shown in Table 7.

Pseudo-One-Dimensional Treatment

The results of the last section taken by themselves do not provide a one-to-one correlation with the hole parameters. An attempt is made here to determine the hole radii directly by means of an approximation which reduces the problem to a one-dimensional one and permits the direct use of the WKB integral.

To do this, the barrier is first sliced into (for the moment arbitrary) elements, with the cuts orthogonal to the metal surface. The probability density of an electron parallel to the surface is constant on the metal (cathode) side of the barrier, so that the amplitude in front of each barrier element is proportional to its base area. It is now assumed that the total probability amplitude on the anode side of the barrier can be obtained by summing the anode-side amplitudes for each element. It is further assumed that the latter are given by the corresponding cathode-side amplitudes multiplied by the appropriate tunneling coefficient D_i, that is,

$$D = \sum (s_i/s)D_i, \tag{22}$$

where s is the total and s_i the base area of the ith element.

The elements are now chosen to be prisms with base areas equal to the hole cross sections. If the inert-gas layers are taken to be crystalline for these purposes, there will be only a limited number of different elements. Furthermore, only the type containing the largest number of holes will contribute appreciably because of the strong exponential dependence of transmission on barrier opacity. The other elements will show up in Eq. (22) with zero weight and will seem to reduce the effective emitting area. This justifies to some extent the assumption of fixed phases implicit in this approximation.

If the holes are taken to be square wells of square cross section and of width $2r_0$ the problem becomes one-dimensional. The penetration coefficient for a barrier slice containing holes can then be found by subtracting the potential profiles of the latter from the WKB integral for the solid slice. This is tantamount to the assumption that the wave function in each hole is an unattenuated oscillatory one, which so to speak transfers the decaying exponential entering the hole from the cathode side unaltered to the exit side.

The deviation of the hole potential from that of a square well is relatively unimportant here, as long as its bottom lies below the Fermi level, that is, as long as the hole cuts completely through the barrier profile (Fig. 100). The detailed hole shape becomes important only if an attempt is made to calculate the reflection of the wave function at each hole, assumed to be zero in the present approximation.

The actual calculation is quite simple. The WKB integral for the barrier penetration of an electron with total energy $-\phi$ (relative to field-free vacuum) is proportional to an area A given by

$$A = \int_0^l (V - E)^{\frac{1}{2}}\, dx = \tfrac{2}{3}K\phi^{\frac{3}{2}}\alpha/F. \tag{23}$$

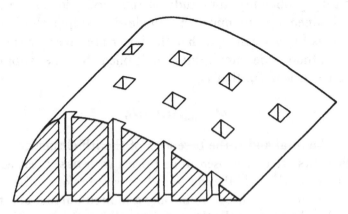

Fig. 100. Schematic diagram of potential barrier for tunneling electrons with discrete wells.

The appropriate A for the hole-containing slice is thus

$$A = \tfrac{2}{3}K\alpha\phi_{\text{mono}}'^{\frac{3}{2}}/F_{\text{vac}} - 2r_0 \sum_j V_j^{\frac{1}{2}}(x_j), \tag{24}$$

where

$$\phi_{\text{mono}}' = \phi_{\text{mono}} - \Delta\phi_{\text{pol}} \tag{25}$$

and $V(x_j)$ is the barrier height at the midpoint x_j of the jth hole:

$$V(x_j) = \phi_{\text{mono}}' - Fx_j/K - 3.6 \times 10^{-8}/Kx_j. \tag{26}$$

When the right-hand member of Eq. (24) is equated with the corresponding expression in terms of the equivalent ϕ_M, $(\tfrac{2}{3})\alpha K \phi_M^{\frac{3}{2}}/F$, the following equation for r_0 results:

$$r_0 = \frac{K\alpha(\phi_{\text{mono}}'^{\frac{3}{2}} - \phi_M^{\frac{3}{2}})}{3F_{\text{vac}} \sum_j V^{\frac{1}{2}}(x_j)}. \tag{27}$$

The values of x_j to be used in Eqs. (26) and (27) are found from the crystal radii of the inert gases and have been chosen for the separation between 100 planes. The value of x_1 has been chosen to be

$$x_1 = \tfrac{1}{2}(d_{\text{W}} + d_{\text{G}}), \tag{28}$$

where d_{W} and d_{G} refer to the crystal diameters of tungsten and the gases respectively.

The results of the calculations are shown in Table 7. The effective r_0 are seen to be considerably less than the corresponding crystal dimensions. The calculations are only slightly altered for other reasonable hole spacings (not shown in Table 7). Consequently the use of Fermi's treatment seems justified. In the present calculation the zigzagging of electrons in order to sample more holes can also be neglected, since the ratio $2r_0/d_w$ is sufficiently small to make the total path under the barrier unprofitably long in such a process.

Matching of the Approximations

It is interesting to compare the results of the two approaches by assuming a hole shape, finding an expression for a from its parameters, and forcing a match with the a found by the Fermi approximation.

The matching of the function $u(r)$ and its first derivative at the hole edge ($r = r_0$) leads to the result

$$a = u(r_0)/u'(r_0) - r_0,\tag{29}$$

as inspection of Fig. 99 shows. The discussion of the preceding section indicates that the holes may be considered as square wells of radius r_0 and depth V_0, so that

$$u(r) = A \sin (\eta r),\tag{30}$$

where A is a constant and

$$\eta = (2m/\hbar^2)^{\frac{1}{2}}(V - E)^{\frac{1}{2}} \cong (2mV)^{\frac{1}{2}}/\hbar.\tag{31}$$

Substitution in Eq. (29) yields

$$a = r_0\left(\frac{\tan \eta r_0}{\eta r_0} - 1\right).\tag{32}$$

If the well depth V_0 is chosen as

$$V_0 = Ze^2/r_0,\tag{33}$$

Z being the effective nuclear charge on the atom, a becomes

$$a = r_0\left(\frac{\tan \gamma}{\gamma} - 1\right).\tag{34}$$

where

$$\gamma = (2Zr_0/r_{\mathrm{H}})^{\frac{1}{2}}\tag{35}$$

and r_{H} is the first Bohr radius of hydrogen; γ is a measure of hole strength and is related to the quantity β occurring in the work of Allis and Morse [3] on the scattering of electrons by gases:

$$\gamma = 2\beta.\tag{36}$$

The value of Z to be used in γ depends on the penetration of the scattered electron and thus on the effective r_0. Equations (34) and (35) will therefore be used to obtain Z from the previously found values of a and r_0. These will then be compared with the results of Allis and Morse on the scattering of slow electrons in inert gases.

The present determination of Z is not completely unequivocal, since Eq. (34) has repeated roots, so that

$$Z = (r_{\mathrm{H}}/2r_0)(\gamma_0 + n\pi)^2,\tag{37}$$

γ_0 being the smallest root of Eq. (34). However, it is possible to make the correct choices by using in Eq. (37) the values of γ lying closest to the corresponding values of 2β of Allis and Morse. In some cases the choices are so obvious that this procedure is unnecessary.

Table 8 summarizes the various results and compares them with

TABLE 8. Comparison of present results with data of Allis and Morse.[a]

Gas	Atomic number	Effective nuclear charge		$\beta [= \frac{1}{2}\gamma]$		Hole radius, r_0 (A)	
		Gomer	A & M	Gomer	A & M	Gomer	A & M
Neon	10	7.9	8.3	2.06	1.71	0.60	0.37
Argon	18	9.1	10.6	2.1	2.7	0.52	0.74
Krypton	36	15.0	15.7	3.8	3.66	1.00	0.90
Xenon	54	22.0	[22.0]	5.3	[4.7]	1.38	[1.06]

[a] The subscript AM indicates the results of Allis and Morse;[3] the other results are those of the present work; the AM data on xenon have been found by extrapolation.

those of Allis and Morse. These authors used a cut-off Coulomb potential in their calculations. This does not differ appreciably from a square well in the region of interest. It is seen that the agreement of the Z, r_0, and γ values is quite good, considering the relative crudity of our approximations. It can be concluded that the mechanism proposed here is essentially correct and that the approximations made in evaluating it are reasonable.

The most noteworthy result is the conclusion that electrons tunnel *through* rather than between atoms in this case.

MOLECULAR IMAGES

A field-emission experiment first performed by E. W. Müller in 1950, and possibly the most widely known, consists in subliming small amounts of copper phthalocyanine or other large molecules onto a tungsten emitter.[4] The startling result is the presence of small, bright images on the main pattern (Figs. 101 and 102). These consist of doublets or quadruplets of bright spots and can be seen to undergo movement as a unit if the gas pressure is higher than 10^{-8} mm-of-mercury. Under these conditions a fourfold "clover-leaf" occasionally changes to a doublet and vice versa.

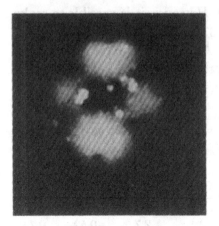

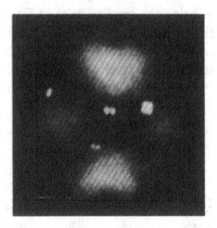

Fig. 101. Field-emission pattern from a tungsten tip covered with a few molecules of zinc phthalocyanine.

Fig. 102. Same tip with more molecules.

The structure of phthalocyanine is that of a flat platelet with fourfold symmetry and investigators were therefore tempted to interpret the patterns as images of single phthalocyanine molecules, showing four corners for flat and two for "standing" ones. Unfortunately, it turned out that both two- and fourfold symmetric patterns could be obtained with a large number of molecules regardless of their shape or symmetry.[4] Thus even threefold symmetric molecules show two- and fourfold symmetric patterns. Hence there is no straightforward correlation between molecular shape and image shape.

The phenomenon is intriguing and it is interesting to speculate on its mechanism. First, the known experimental facts will be summarized.

(1) The evaporation of a variety of substances onto field emitters results in the appearance of small, usually bright, patterns, or images, distributed somewhat at random, but with a noticeable preference for the close-packed planes of the substrate or their vicinals (Figs. 101 and 102). Experiments by Melmed and Müller[4] indicate that an image appears for every 2–15 molecules hitting the tip, depending on the compound tested. This indicates either a moderately low sticking coefficient, some degree of polymerization in the vapor, or the fact that not every adsorbed molecule gives rise to an image. There is some evidence that all of these effects are operative to some extent.

(2) The electron images correspond to an apparent size of 100–

200 A, based on the over-all magnification. Field-ion patterns often show much smaller bright spots where images appear on the electron patterns.[5] Thus it seems that local magnification, scattering, or diffraction are responsible for the size of the electron images.

(3) The brightness of images relative to the substrate depends on the ionization potential of the molecules and the work function of the substrate. Thus flavanthrene and copper phthalocyanine show up dark on barium-covered tungsten ($\phi = 2.5$ ev).[4] Similarly, oxygen images are seen only on oxygen-covered tips (and then only rarely) but not on clean ones. This suggests that field emission occurs, at least under some conditions, from the ad-molecules themselves.

(4) Images are observed to rotate, to librate, to change from doublets to quadruplets and vice versa, particularly when the residual gas pressure in the tube is high. Desorption or movement of only a portion of an image is never observed. However, changes in size and intensity have been noted. Müller and Melmed suggest that these may be due to stepwise desorption of individual molecules from a small stack.[4]

(5) Increases in field can cause slight increases in the size and separation of the spots in doublets or quadruplets.

It is clear from the foregoing discussion that very small aggregates, in many instances almost certainly individual molecules, can give rise to local enhancement and divergence of emission. The gross features of the phenomenon can be explained by two different mechanisms. The first [6] applies when the adsorbed molecule is large and has a sufficient number of quasi-free (for example, π) electrons or at least levels to keep the field out of its interior. In that case the adsorbate will simulate a metal, act like a conducting protrusion on the tip surface, and cause local field enhancement and the lens effect discussed in Chapter 2.

The presence of a sufficient number of quasi-free electrons in such a molecule also implies that the energy levels are fairly closely spaced and that there are empty ones available. Consequently, the highest filled state will equalize itself with the Fermi level of the substrate by the requisite amount of electron transfer to or from the molecule. Under these conditions emission will occur from the adsorbed molecule and will be governed by its work function, that is, its ionization potential (within the level spacing), and by the local field, Eq. (2.5) (Fig. 103).

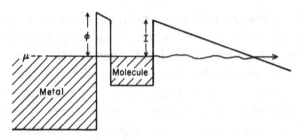

Fig. 103. Schematic potential-energy diagram explaining enhanced emission from an adsorbed molecule; no field penetration; μ, Fermi level; ϕ, work function of metal; I, ionization potential of molecule.

The emission from single molecules of zinc phthalocyanine has been measured as a function of applied voltage and interpreted in terms of this mechanism. If field enhancement is ignored, an apparent ionization potential of 4.2 ev is found. Electrolytic trough measurements on molecular models indicate enhancement factors of 2.5 at the edge and 1.5 near the center of the zinc-phthalocyanine molecule. The former value leads to an ionization potential of 7.5 ev, the latter to 5.2 ev.[6] Measurements by Melmed [5] on the hydrogen-ion image-appearance voltages of zinc-phthalocyanine molecules relative to the neighboring substrate indicate that the lower value of the enhancement factor and hence of the ionization potential of zinc-phthalocyanine is more nearly correct.

The emission from a phthalocyanine molecule can be calculated with this model by multiplying the impingement rate of π electrons at the outer surface by the appropriate penetration coefficient.[6] The result agrees very well with the experimentally found value, 10^{-10}–10^{-9} amp. It can also be shown [6] that the supply of electrons arriving at the inner surface of the molecule from the metal exceeds the emitted current by at least 10,[6] so that the emission can be regarded as coming from the molecule itself. It therefore appears that the assumed mechanism is correct for molecules with a large number of relatively free electrons.

The second mechanism [4] corresponds to complete field penetration through the molecule and is very similar to that advanced for emission through layers of inert-gas atoms in the last section. The molecule constitutes a short-range potential well and thus acts as a window in the barrier both in field emission and field ionization (Fig. 104). On the other hand, polarization leads to a local decrease in field, or,

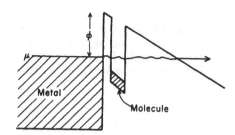

Fig. 104. Potential-energy diagram for an adsorbed molecule in the presence of a high field; complete field penetration assumed. The molecule acts as a window in the potential barrier.

more accurately, to the induction of an unfavorable dipole moment. Enhanced emission can occur when the first effect outweighs the second. Magnification results from a distortion of the lines of force in the molecule (acting as a dielectric pill) or from scattering by the potential hole. This mechanism is most likely to apply in cases where a molecule does not have a large number of quasi-free electrons to prevent field penetration.

These cases have been presented as distinctly different largely for greater clarity of presentation. It is clear that some overlap will exist in almost all cases. Thus it is quite likely that some electrons will tunnel directly from the metal even in the first case, where the field does not penetrate the molecule. In addition, there will always be some field penetration because of the fact that molecules do not really have surface states. The surface charge density required to exclude a field F v/cm is $F/1200\pi$ esu/cm^2. For $F = 3.10^7$ v/cm, this corresponds to 2.10^{13} electrons/cm^2 or 0.2 electron per phthalocyanine molecule.

The following naive argument indicates that phthalocyanine molecules can readily supply even larger charge densities. A field F induces a dipole moment αF in a molecule of polarizability α. If the molecule is considered to be a flat box of depth d and the dipole moment to arise from the distribution of surface charges on the pair of faces normal to the field, the charge induced on each of these faces is $\alpha F/d$. For phthalocyanine, $\alpha = 1.2 \times 10^{-22}$ cm^3, so that for $F = 3.10^7$ v/cm and $d = 3$ A, $q = 4.10^{-10}$ esu or approximately 1 electronic charge, which is ample. However, this cannot be strictly a surface charge when the molecule is adsorbed flat, so that the equipotential surface corresponding to the substrate will always enter the molecule to some extent. Increases in field beyond a certain point will lead to more field penetration and a greater deformation of the molecule potential

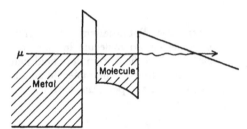

Fig. 105. The system depicted in Fig. 103 with partial field penetration; this results in a lowered effective work function for the molecules.

(Fig. 105). If the highest filled state in the latter corresponds to the Fermi level, which remains constant, a decrease in the effective work function results. A mechanism of this kind was proposed by E. W. Müller some years ago.[7] It is completely analogous to field penetration in semiconductors and is discussed in detail for that case in Chapter 1 (p. 25).

The fact that ion images of molecules are generally smaller than the corresponding electron patterns deserves discussion, since it might seem to argue against field enhancement and a lens effect. However, the lines of force emerging from a molecule are compressed after traveling normal to its surface for a very short distance. Consequently, electrons attempting to follow these lines from their beginning will have more divergent trajectories than ions formed at least 5 A from the surface where the lines of force have already become compressed. Another way of saying this is that the distortion of the equipotentials has been largely damped out at the ion-forming distance.

It is very difficult to explain the observed image shapes in detail. These could be scattering patterns [4] or possibly they correspond to the momentum distribution of the emitted electrons.[6] The absence of threefold patterns for molecules with that symmetry is extraordinarily hard to account for on any basis, except that of polymerization.

It is clear that many aspects of this phenomenon are still mysterious and will require further effort.

FIELD EMISSION FROM WHISKERS

It was pointed out in Chapter 2 that the general applicability of field-emission techniques is severely limited by the relatively small number of materials from which successful emitters can be fabricated by conventional means. For this reason alone field emission from whiskers would be interesting, since it greatly extends the range of emitter materials. In addition, a great deal can be learned about the properties and growth mechanisms of whiskers by such studies, as the following discussion will show.

Mercury Whiskers

In 1953 Sears [8] noted that mercury vapor from a reservoir at $-40°$ to $-20°$ C did not condense in bulk on glass surfaces at $-60°$ to $-80°$ C, but formed a number of very thin filamentary growths or whiskers. These could be seen only with scattered light and underwent Brownian motion. Sears was able to estimate their thickness as 200 A from the amplitude of this motion and reasonable assumptions on the elastic moduli of mercury.

Sears postulated the following growth mechanism. At low supersaturation the formation of hypercritical nuclei is difficult or impossible because reevaporation exceeds growth. (The high vapor pressure of nuclei is caused by their small radii of curvature, in accordance with the Gibbs-Thomson equation.) However, certain spots on a surface may adsorb mercury particularly well, resulting in formation of stable nuclei. If these happen to contain a screw dislocation, axial growth without thickening can occur, since the dislocation provides

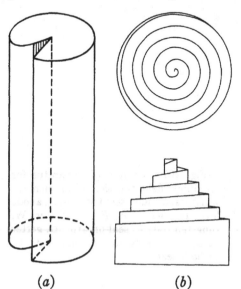

Fig. 106. Schematic diagram showing an axial screw dislocation in a cylindrical crystal: (a) dislocation before axial growth; (b) crystal end face after growth has proceeded.

(a)   (b)

a preferential anchorage site which cannot be cancelled by condensation (Fig. 106). Sears also showed that the impingement rate on the ends of the whiskers was quite insufficient to account for the estimated growth rates, and assumed that surface diffusion over the sides of the whiskers supplied the necessary flux.

The thinness of these whiskers suggested their use as field emitters.[9] Success hinges on two factors. They must be sufficiently strong to withstand the stresses caused by the field and they must be anchored equally strongly to the substrate. As it happens, both requirements are met.

The apparatus used for growth and emission *in situ* is shown in Fig. 107. A small tungsten wire sealed into the cold finger, ground off, and electropolished provides a conducting substrate of controllable size.

Growth occurs readily on the tungsten substrate and there is no difficulty in obtaining field emission at moderate voltages. Typical patterns are shown in Figs. 108–110. Usually only a few whiskers

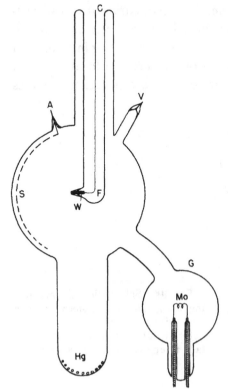

Fig. 107. Diagram of apparatus for obtaining field emission from mercury whiskers: S, fluorescent screen; A, anode lead; Hg, reservoir; F, cold finger; W, tungsten rod; V, seal-off arm; G, getter bulb; Mo, molybdenum filament; C, cathode lead.

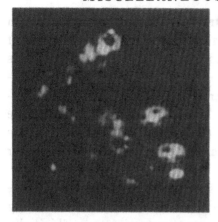

Fig. 108. Field-emission patterns from
several mercury whiskers.

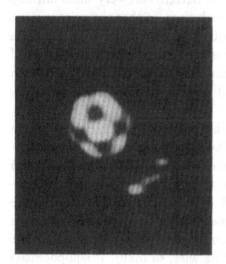

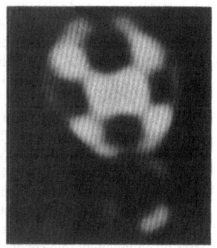

Figs. 109 and 110. Field-emission patterns from single, 110 oriented
mercury whiskers.

appear at a given voltage so that there is little overlap between
patterns. These correspond to single crystals of 110 orientation (fcc
pseudo system). The fact that definite crystal faces appear, as is
clearly shown by the observed symmetry, means that the whisker tip
must be curved, for example, spheroidal, and not flat. This would be
expected if there is a central screw dislocation, but would also result
because of surface tension.

Emission currents of 10^{-9}–10^{-8} amp can be drawn from single
whiskers. It will be seen later that this corresponds to current densities

of the order of 10^2–10^3 amp/cm^2. Insertion in the Fowler-Nordheim equation with $\phi = 4$ ev indicates that this corresponds to fields of 3.10^7 v/cm. Equation (2.34) therefore shows that mercury whiskers are able to withstand tensile stresses of $\sim 10^9$ dy/cm^2. This remarkable strength indicates that whiskers are almost free from imperfections and so lends weight to Sears's mechanism, which implies the absence of imperfections other than the axial screw dislocation. It has been shown by Eshelby [10] that the latter need not weaken the whisker appreciably.

Since the pressures required for growth are of the order of 10^{-5}–10^{-6} mm-of-mercury, it is possible to observe the whiskers during the process. It is then noticed that the patterns are very small initially but increase in size and brightness with time, so that the voltage must be reduced continuously to prevent excessive emission. Increases in pattern size can arise only from lessened compression of the lines of force at the whisker tip. This indicates that the changes result from the increase in whisker length, which lessens the effect of the substrate on the field at the end.

This electrostatic problem can be treated by elementary means with only minor simplifications.[9] The system is regarded as a concentric spherical condenser with a whisker of radius r_w and length h growing from the inner sphere, whose radius is r_s. Since we are interested primarily in the variation of F with h and not in its absolute value, it is permissible to replace the whisker by a spherelet of radius r_w, a distance h from the substrate surface, but connected to the latter by an infinitely thin conducting wire, so that it is equipotential with it. The problem is reduced to that of determining the charge required at the center of the spherelet to make its surface equipotential with the substrate and then finding the field due to this charge.

Since the inner sphere of radius r_s is very much smaller than the outer one, the potential V at a distance r from its center is q/r, where q is given by

$$q = r_s V(r_s) \equiv r_s V_0. \tag{38}$$

The potential at the spherelet surface before electrical connection is

$$V(r_s + h) \cong q/(r_s + h), \tag{39}$$

since its radius is very small compared with r_s and h. A charge

$$q' = r_w[q/r_s - q/(r_s + h)] \tag{40}$$

must therefore be placed at its center to bring the potential to V_0. The field at the spherelet surface is then

$$
\begin{aligned}
F &= q'/r_w{}^2 \\
&= (q/r_w)h/r_s(h + r_s) \\
&= (V_0/r_w)h/(r_s + h).
\end{aligned}
\tag{41}
$$

Since the field at a free spherelet of radius r_w charged to potential V_0 is

$$
F_{\text{free}}^{\text{sphere}} = V_0/r_w,
\tag{42}
$$

Eq. (41) becomes

$$
F/F_{\text{free}} = h/(h + r_s).
\tag{43a}
$$

If the substrate is a cylindrical wire instead of a sphere, F/F_{free} takes an even simpler form

$$
F/F_{\text{free}} \propto \ln(1 + h/r_s)
\tag{43b}
$$
$$
\widetilde{\propto}\ h/r_s \quad \text{if } h \ll r_s.
$$

As expected, this expression does not contain the whisker radius explicitly. Since F is fixed at $\sim 3.10^7$ v for constant emission and F_{free} is proportional to the applied voltage V_0, Eq. (43a) can be used to convert voltage-time into length-time curves. Rearrangement and some manipulation yield

$$
h(t) = h(0)\frac{V(0) - V(\infty)}{V(t) - V(\infty)} \cong h(0)\frac{V(0)}{V(t)},
\tag{44}
$$

with

$$
V(\infty) \cong 5r_w F_{\text{free}}^{\text{whisker}} \cong 5r_w \times 3 \times 10^7,
\tag{45}
$$

where all quantities are considered as functions of time and $t = 0$ refers to the start of the measurements. Figure 111 shows a voltage-time curve and Fig. 112 curves of log $[h(t)/h(0)]$ versus time. It is seen that growth is positively exponential until it suddenly levels off and almost stops.

This shows that growth occurs by impingement of mercury atoms on the whisker surface and diffusion to the growing end. If diffusion is sufficiently rapid, the controlling step will be the rate of impingement. Under these conditions the growth rate is given by

$$
\frac{dh}{dt} = \frac{2\pi r_w h P m \alpha}{\pi r_w{}^2 \rho (2\pi m k T)^{\frac{1}{2}}} = \gamma h,
\tag{46}
$$

where

$$
\gamma = 6.96 \times 10^{-3} P(\alpha/r_w)\ \text{sec}^{-1}.
\tag{47}
$$

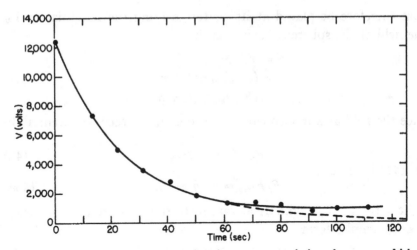

Fig. 111. Curve of voltage versus time for constant emission of mercury whiskers during growth. Broken curve indicates the curve predicted by the exponential growth law, as $t \rightarrow \infty$.

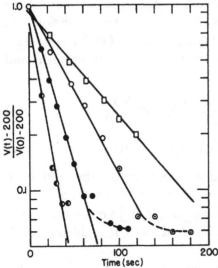

Fig. 112. Semilogarithmic graph of $[V(t) - V(\infty)]/[V(0) - V(\infty)]$ versus time for mercury whisker growth at various pressures; $V(\infty) = 200$ v in this case. Note leveling off of curves.

In these expressions P (mm-of-mercury) is the vapor pressure of mercury, ρ the density of the whisker, and α the sticking coefficient of impinging Hg atoms. It is possible to find r_w/α from the pressure variation of the growth constant γ and hence to obtain an upper limit on r_w. This turns out to be 70–100 A, in remarkable agreement with Sears's value. Since r_w cannot be appreciably smaller than this without resulting in the absence of definite crystal faces in the observed patterns, α must be close to unity.

Exponential growth ceases when an atom landing at the whisker base can no longer catch up with the growing end or evaporates before it gets there. The first condition is equivalent to stating that for exponential growth

$$(D\tau)^{\frac{1}{2}} \geqslant he^{\gamma\tau}, \tag{48}$$

where τ is a critical time, as yet unknown. Squaring both members of Eq. (48) and introducing the new variable

$$x = 2\gamma\tau \tag{49}$$

leads to

$$\frac{Dx}{2h^2\gamma} \geqslant e^x. \tag{50}$$

The left-hand member of Eq. (50) is a linear function of x, the right-hand member an exponential one. Equality corresponds to an atom's just catching up with the growing end, and implies tangency of the linear and exponential functions of x, that is,

$$D/2\gamma h^2 = e^x. \tag{51}$$

Equations (50) and (51) can hold simultaneously only if $x = 1$, or if

$$\tau = 1/2\gamma. \tag{52}$$

Substitution in Eq. (51) and solution for D shows that

$$D \geqslant 5.44\gamma h^2 \tag{53}$$

for exponential growth.

Since it is possible to estimate terminal lengths either by direct microscopic examination under scattered light or electrically from Eqs. (43) and (45) and the values of r_w and r_s, Eq. (53) can be used to estimate the surface diffusion coefficient of mercury on mercury. This turns out to be 5.4×10^{-5} cm^2/sec at $-78°$ C with terminal lengths of $h = 0.01$ cm. Use of Eq. (2) for the diffusion coefficient gives an estimate of ~ 1.1 kcal for the upper limit of E_d. Growth experiments can be performed at various substrate temperatures, so that it should be possible to obtain a much more accurate value of this quantity from the temperature variation of the terminal length.

Under conditions where re-evaporation of impinged atoms terminates growth, Eq. (4.8) applies, so that a *lower* limit of $E_{ad} \sim 10$ kcal results. Since the heat of sublimation of mercury is 15 kcal, this is reasonable.

Although the conditions for exponential growth are unequivocal, the fate of the whisker after its cessation is not. If diffusion is limiting, the concentration of adsorbed Hg atoms on the whisker shank will increase. If new layers can be nucleated with this slight increase in supersaturation, these will grow very rapidly to the length of the whisker, thereby competing successfully with the screw dislocation for impinging atoms. Under these conditions axial growth will stop and thickening will take its place until no new layers can be formed. Axial growth may then continue. If the formation of new layers is difficult in the first place, exponential will be followed by linear growth with a rate numerically equal to the terminal exponential value, until the supersaturation on the shank increases to the point where new layers can be formed. Thickening will then occur again, and so on. It should be noted that a very high degree of perfection can be expected only for the initial exponential stage of development.

The presence of screw dislocations in the whiskers is made very plausible by the arguments just presented. Even more direct evidence comes from the observation that whiskers sometimes twist elastically when the field and hence the stress is increased. This strongly suggests a built-in spiral dislocation, although a quantitative evaluation is difficult.[11]

Occasionally it is possible to pull whiskers off piecemeal. This indicates extremely strong adhesion to the substrate. Experiments have shown [9b] that mercury whiskers can be grown and used for field emission on a large variety of substrates, such as aquadag, conducting glass, aluminum foil, and so on. The 110 orientation is preserved in each case. It was also found that whiskers will *not* grow on a really smooth surface, such as tungsten heated to 2000° K and kept either clean or covered with a monolayer of oxygen. This suggests very strongly that whiskers grow in the cracks present on most surfaces so that their strong adhesion comes from lateral rather than basal contact. This may also explain the origin of screw dislocations. If two nuclei growing on opposite walls of a crack grow together with some mismatch, the resulting dislocation could easily have a screw component. The almost invariably observed 110 orientation of the whiskers is probably due to the fact that a random dislocation is most likely to have a large screw component along this direction. Also, the faces parallel to 110 constituting the whisker sides are close-packed and therefore permit sufficiently rapid diffusion of impinged atoms.

Experiments on Other Whiskers [12]

The feasibility of using mercury whiskers as field emitters suggests that almost any other material can be utilized in this way. This is the case. The apparatus consists of a field-emission tube containing a heatable tungsten wire substrate surrounded by a heatable tungsten loop wrapped with the material to be evaporated (Fig. 113). In electron or ion emission this loop serves as an electrode ring, equipotential with the screen.

If the substrate consists of a twisted rope of very fine tungsten wires (diameter 0.0005 in.) it is possible to clean it by high-temperature flashing without loss of nucleation sites, which occur at the points of contact between strands of the rope. Since the whiskers are thus anchored to large areas of clean metal, the strength of the whisker-substrate bond is enormously increased over that provided by cracks in an oxide layer. Thus it has been possible to field-desorb nickel from its own lattice in a whisker, without tearing the latter from its base.

Emission has been obtained from a large number of materials by this technique. A partial list includes Va, Ti, Pt, Ni, Fe, Cu, Au, Ag, Al, and Ge. There is no reason to believe that the method cannot be

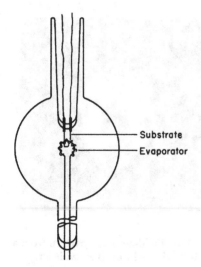

Substrate

Evaporator

Fig. 113. Apparatus for growth and field emission from metal whiskers.

applied to any conducting material. Figures 114–120 show the appearance of typical emission patterns. It will be seen that these are often indistinguishable from "ordinary" field-emission pictures, but that occasionally faces seem less fully developed than in such pictures. This must be ascribed to the small whisker radii and occasionally to build-up caused by resistive heating of the whiskers by the emission current.

Although it is not possible to determine whisker radii from growth rates unless the effective pressure of metal vapor is known, estimates

Fig. 114. Field-emission pattern from a 110 oriented aluminum (fcc) whisker.

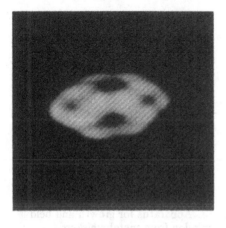

Fig. 115. Field-emission pattern from a 110 oriented gold (fcc) whisker.

Fig. 116. Field-emission pattern from a 100 oriented iron (fcc) whisker.

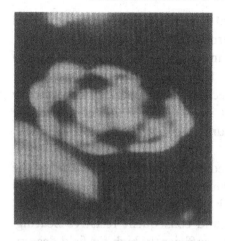

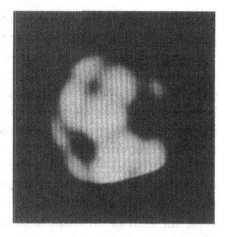

Fig. 117. Field-emission pattern from a 100 oriented iron (bcc) whisker.

Fig. 118. Field-emission pattern from a 110 oriented copper (fcc) whisker.

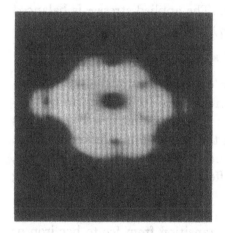

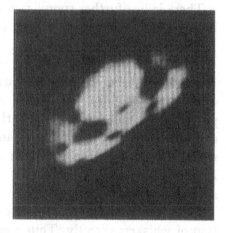

Fig. 119. Field-emission pattern from a 110 oriented titanium (bcc) whisker.

Fig. 120. Field-emission pattern from a 211 oriented vanadium (bcc) whisker.

are possible by comparing the relative size of molecular images on whiskers and on tungsten tips of accurately known radius. As Eq. (2.3) indicates, the additional magnification of small bumps is a function of the ratio of substrate and bump radii, so that in the comparison with tungsten tips the effective molecule radius cancels out. It was found in this way that the radii of metal whiskers generally fall in the range of 50–150 A.

At the time of this writing several interesting applications directed primarily to the study of whisker properties have been noted.

Kinetics. It has been possible to investigate the growth kinetics of whiskers other than mercury by the electric method previously described. It appears that the mechanism postulated for mercury holds also for other metals, at least in the early stages of growth. It should be possible to investigate surface self-diffusion coefficients in this way. (See Appendix 2.)

Tensile Testing. As already pointed out, it is possible to anchor the whiskers very firmly and to determine their rupturing strength by finding the field at which they break. To do this it is necessary to reverse the field to prevent emission and consequent resistive heating. Since many whiskers withstand fields sufficiently high for field desorption from the lattice, it may be necessary to apply the field in short pulses.

There is one further complication. The applied stress τ is balanced both by bulk and surface-tension (γ) forces:

$$\sigma_{\text{appl}} = (F/300)^2/8\pi = \gamma/r + \sigma_{\text{bulk}}. \tag{54}$$

If a representative value of 1000 dy/cm is chosen for the surface tension γ, it is seen that an applied stress of $\sim 10^9$ dy/cm can be supported by surface forces alone if the whisker radius $r \sim 100$ A. Thus a sizable error can be introduced in the case of thin whiskers of weak materials. On the other hand, in the case of very weak substances, it may be possible to turn this effect to advantage by measuring surface tensions in this way.

Phase Transformations. It is often possible to watch the transformation of whiskers directly. Thus the transition from fcc to bcc iron or from bcc to hexagonal titanium and so on can be observed.

Effect of Surface Energy on Structure. It is often noted that conventionally unstable phases persist and even predominate. Thus the fcc iron and bcc titanium structures are frequently observed at room temperature. This indicates that surface energy is able to influence not only crystal habit but also crystal structure when the surface-to-volume ratio is very high. It should thus be possible to investigate the existence of new phases.

A very curious crystal form results [13] when fcc whiskers are grown in ambient pressures of 10^{-6}–10^{-4} mm-of-mercury of oxygen or water vapor (Fig. 121). The structure has fivefold rotational symmetry and consists of five twins with a common 110 axis along the whisker direction. Figure 122 shows a model of this structure. Its sides consist of 100 faces, while the tip consists of five 111 faces, in agreement with Fig. 121. It is probable that this results from a marked change in surface-tension anisotropy by adsorption, making 100 the face of lowest energy. Since the internal twin boundaries are of very low energy, the fivefold structure has only low-energy external and internal

Fig. 121. Field-emission pattern from a nickel whisker, showing fivefold symmetry.

Fig. 122. Cork ball model of an (fcc) structure composed of five twins arranged about their common 110 axis: (a) top view; (b) side view. The sides consist of 100 faces, the tips of 111 faces.

surfaces. The kinetics of the phenomenon are not understood at present, although it seems likely that a poisoning of the screw dislocation of normal whiskers may be involved.

Considered from the viewpoint of field emission the use of whiskers grown *in situ* has two advantages which greatly extend the range of substances accessible to study. First, whiskers are very strong, since they are nearly perfect, and can be made to adhere strongly to clean substrates. Second, they can be grown under high-vacuum conditions. In this way materials that could never be cleaned by conventional means can be made into clean emitters. The method also eliminates the difficulties encountered with involatile impurities diffusing to emitter surfaces.

In their present stage of development, whisker emitters suffer from two minor disadvantages. First, it is not always possible to obtain emitters singly. However, a suitable choice of substrate geometry makes it possible to circumvent this where necessary. Second, emitters are generally very thin. This can be troublesome when fully developed high-energy faces are desired and can lead to some resistive heating by the emission current. If the resistivity ρ depends linearly on temperature t,

$$\rho = \rho_0 t + a = \rho_0 t', \tag{55}$$

where

$$t' = t + a/\rho_0 \tag{56}$$

and a and ρ_0 are constants, the temperature at the whisker end h can be found from the relation

$$t'(h) = t'(0)/\cos(\beta h), \tag{57}$$

where

$$\beta = i(\rho_0/K)^{\frac{1}{2}}, \tag{58}$$

i is the current density in the whisker and K its thermal conductivity.

It is probable that these difficulties can be overcome, at least in part, by a proper choice of growth conditions.

Appendix 1

Although the construction and use of field-emission tubes is hardly more complicated than that of other cathode-ray devices, a certain amount of specialized experimental knowledge has developed in the last few years. Since this information is scattered through the literature [1,2] or not in print, it is worthwhile to discuss it briefly.

Construction of Tubes

The shape of the tube is usually a matter of personal taste or convenience. Spherical tubes half or fully coated with screens are simplest to construct. For ion work, where photography at low intensities necessitates the use of large-aperture objectives, the resultant small depth of focus makes the use of flat screens necessary. For most purposes [3] tubes are constructed from Pyrex glass, since this is easy to handle and can be baked out readily.

Screens are made from standard, commercially available phosphors.[4] For work at room temperature blue zinc sulfide has the highest light yield, although it is no brighter than willemite at 4° K. Zinc sulfide has the advantage of being very fine grained, but it is chemically sensitive and can be altered by evaporated metals, by excessive heating, particularly in air, and by ion bombardment. Willemite is considerably more stable and only slightly less bright, although generally green rather than blue.

There are three commonly used methods of screen deposition.

(1) *Decantation.*[4] A suspension of finely ground phosphor in water containing small amounts of sodium silicate is poured into the tube and the phosphor is allowed to settle out. The tube is then tilted at a rate not exceeding 1 rev/hr and the liquid is decanted. This method

can be used only with nearly flat screens and while widely employed in industry is not very suitable for small-scale applications.

(2) *Nitrocellulose Method.*[4] A suspension of phosphor in nitrocellulose–amyl acetate solution [4] with a drop of butyl phthalate as plasticizer is poured into the tube and rotated by hand until a smooth film of lacquer has covered all the desired portions. This is dried by gently pulling air through the tube. Slow heating to 400° C (overnight) decomposes the nitrocellulose gently and leaves the phosphor in place. Excess can then be removed by wiping. Further heating to 450–550° C stabilizes the phosphor. The method is therefore best for willemite. While very uniform screens can be made in this way, the technique requires a certain amount of practice. The chief danger is formation of gas bubbles during the decomposition of the lacquer.

(3) *Phosphoric Acid Method.*[5] A few drops of concentrated phosphoric acid are dissolved in 1 ml of acetone or methanol. One to five drops of this solution are pipetted into the tube and spread evenly by introducing 10–20 small glass spheres (1–2 mm in diameter) and shaking. After spreading and evaporation of the solvent, the freshly reground phosphor is sprayed into the tube with an atomizer. Excess is removed by tapping, and the rest of the tube is wiped clean with cotton dipped in water. The tube is then heated to 400° C for an hour or two. After this treatment the phosphor cannot be removed even by severe abrasion. This method is by far the simplest and produces quite uniform screens.

It is necessary to make the screen conducting. This can be done in two ways.

(1) *Conducting-Glass Method.*[6] In this method the tube is coated with an optically transparent conducting layer before phosphor deposition. This method is excellent at low and moderate current densities ($\leqslant 10^{-6}$ amp/cm² of screen) where charging of the screen particles is unimportant. The conducting coating is put on by heating the tube to 450–475° C and blowing into it the vapor formed over molten stannous chloride (Fig. 123). Parts that are to be nonconducting are masked with aquadag or some other removable coating. This method is rapid and simple. The coating withstands all acids except hydrofluoric, and is still conducting at 1.3° K.

(2) *Metallic Backing.* This is achieved by evaporating a metal, in vacuum, on the back of the phosphor layer. The backing must be

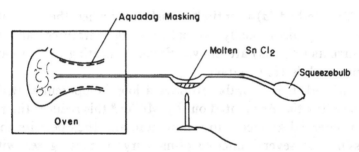

Fig. 123. Schematic diagram of arrangement for producing a transparent conducting coating on glass.

thick enough to appear metallic and reflecting. If the phosphor was deposited as a nitrocellulose lacquer, it is best to evaporate the metal film before its decomposition.[7] A vacuum of 10^{-5}–10^{-6} mm-of-mercury can be attained during evaporation if the outside of the tube is water cooled during the operation. The tube is then opened to air and the nitrocellulose decomposed by heating. If this is carefully done, the metallic backing is not damaged in the process and remains coherent. Backings prepared in this way from platinum or even aluminum are very stable and resistant to water or acetone. Above 5000 v electron absorption is small for aluminum, so that the brightness exceeds that of unbacked phosphors because of light reflection. Below this voltage brighter images can be obtained with unbacked screens.

External Screen Connections

It is necessary to have an external electric lead to the screen. This can be provided in a number of ways. The two most satisfactory consist of tungsten-Nonex seals to insure vacuum tightness and (*a*) a spring contact with a conducting-glass layer (Fig. 124) or (*b*) a collapsed platinum-coated Nonex sleeve on a section of the tungsten rod which has been covered with platinum foil by spotwelding [8]

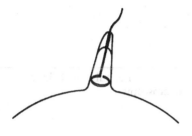

Fig. 124. Schematic diagram of a vacuum-tight electric spring contact.

(Fig. 125). Method (*a*) is entirely satisfactory when the apparatus is subjected only occasionally to bake-out temperatures, and when small currents ($< 1 \mu a$) are drawn. Frequent heating tends to anneal the spring and destroy contact.

For ion work it is desirable to have a loop equipotential with the screen near the tip. As pointed out by Müller,[9] this reduces the region of steep potential gradients to a minimum and thus permits the use of pressures of several microns-of-mercury of most gases without danger of discharge. It is possible to combine this loop with the electric lead to the screen (Fig. 126).

Tip Assemblies

The basic assembly consists of a heatable wire loop to which the tip is spotwelded or on which it grows if whiskers are used. In the latter case it is desirable to reduce the apex of the loop by etching.

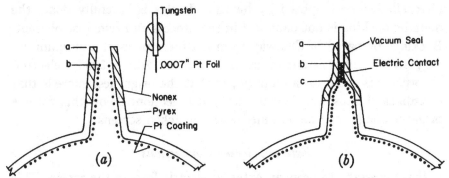

Fig. 125. Diagram showing construction of another type of vacuum-tight electric contact: (*a*) before; (*b*) after sealing.

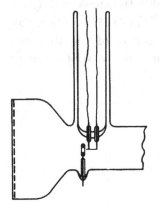

Fig. 126. Schematic diagram of tube for field-ion microscopy.

This not only results in larger patterns, because of lessened compression, but insures nonuniform heating. In this way the central section can be held at the proper growth temperature while the rest of the loop is colder. This insures that whiskers will grow only where they are wanted.

In order to obtain accurate temperature control it is often desirable to spotweld potential leads to the loop near the apex and to determine T from the measured resistance and a previous R-vs.-T calibration.[2,10] The nonuniformity in temperature over the loop is given by [10]

$$t/t_0 = \frac{\cos(\alpha x)}{\cos(\alpha l)}, \tag{1}$$

where t_0 is the temperature at a reference point $x = \pm l$ and t the temperature at the apex, $x = 0$; α is given by

$$\alpha = \frac{Il}{A}(\rho_0/K)^{\frac{1}{2}}, \tag{2}$$

where I is the heating current, A is the cross-sectional area of the wire, K is the thermal conductivity, and ρ_0 is related to the resistivity ρ by the equation

$$\rho = \rho_0 t, \qquad t = T + \text{constant.} \tag{3}$$

In most cases the nonuniformity can be neglected. Corrections must be made for the temperature drop along the tip itself above 1000° K where radiation losses become important. As was pointed out in Chapter 3, it is possible to devise automatic temperature controllers to raise and keep a loop at a given R and hence T.

If tubes are to be used at low temperature (4–20° K), nichrome or other alloy sections must be inserted in the tip assembly,[2,10] as shown in Fig. 127. This is necessary because the thermal and electric conductivity of pure metals decreases by a factor of 100 from the 300° K values at 20° K, so that very small increases in the (large) total current cause sudden sharp temperature rises as the loop regains its resistivity. The insertion of alloys, as thermal barriers, permits smooth temperature control.

Etching of Tips

There are a number of etch methods for different metals. Electrolytic etching at 2–10 v ac in molten sodium nitrite works for a very large

Fig. 127. Tip assembly for low-temperature work: N, nichrome; T, tip; L, heating loop; P, potential leads.

number of materials.[1] Tungsten and molybdenum can be etched very easily at 2–10 v ac in $1N$ sodium hydroxide solution. Nickel, copper, iron, and so forth can be etched in concentrated phosphoric acid at the same voltage. Tantalum must be etched in hydrofluoric-sulfuric acid solution with direct current to avoid hydrogen embrittlement. Experiment will reveal other etch solutions.

In all cases the tip wire is inserted vertically, or as nearly vertically as possible, into the etch bath and is examined periodically under the optical microscope at 300–500X magnification. A correctly etched tip should show a long, slender taper and be unresolvable at the end. As pointed out in Chapter 1, a long taper prevents excessive blunting on heating. With a little experience it is possible to predict from the taper alone what the eventual radius of the tip will be after heating. It is preferable to perform all spotwelds before the final etch and to electropolish the whole assembly in the etch bath, to get rid of spurious emission. This is important if accurate work-function measurements, based on emission currents, are to be made.

Vacuum Systems

The attainment of ultrahigh vacuum is relatively simple if the system can be heated to $\sim 430°$ C. The use of copper-filled traps, originated by Alpert,[11] provides a getter in the vacuum path. On baking out copper redistributes itself, partly by surface diffusion and partly by the formation of volatile compounds which decompose again, so that the foil attains enormous real areas, capable of chemisorbing vast amounts of gas. During bakeout all electrically heatable parts are flashed repeatedly. Oil or mercury diffusion pumps may be used equally well. The use of thin wire collector ion gauges,[11] such as the Westinghouse WL5966 gauge,[12] is advantageous, not only because these measure pressures below 10^{-8} mm-of-mercury reliably but also because they contain less metal than conventional types.

The use of various all-metal valves, such as the Alpert valve,[11] manufactured commercially,[13] is compatible with the vacuum requirements.

For some experiments it is desirable to use permanently sealed-off tubes. High vacuum can be attained in such systems by equipping the tube with an auxiliary getter bulb containing a heatable molybdenum filament. Deposition of a molybdenum film on the walls of this bulb insures pressures of 10^{-10} mm-of-mercury or less if the filament is properly outgassed before seal-off. The glass seal-off arm should be flamed to partial collapse before the final bakeout and seal-off, to minimize gas liberation.

The use of cryogenic techniques is in some ways more complex but has many advantages, chief of which is the fact that bakeout is unnecessary and that high vacuum may be alternated with exposure to gas merely by warming the tube.[2] Details of cryogenic procedures are beyond the scope of this book.[14] A schematic diagram of a cryostat assembly for field-emission work is shown in Fig. 128.

Electrical Equipment

The principal requirements are high-voltage power supplies in the 10–20 kv range for field emission and 20–30 kv range for ion emission. These are available commercially [15] or may be built. For accurate work-function measurements, well-regulated supplies are necessary. This is best achieved by using rectified rf supplies equipped with feedback chains. Voltages can be measured accurately by tapping small sections of high-resistance potentiometers. Accurate current measurement in the 10^{-12}–10^{-6}-amp range is best accomplished with a vibrating-reed [16] or other electrometer.[17]

Figure 129 shows a circuit diagram for a tip temperature controller that has proved very satisfactory. This design is somewhat superseded by transistor circuits.

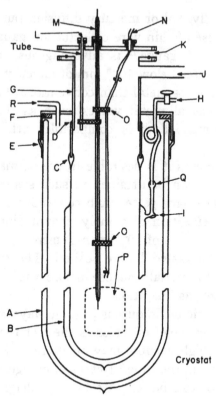

Fig. 128. Diagram of cryostat for low-temperature field-emission work: *A*, liquid-nitrogen Dewar; *B*, liquid helium Dewar; *C*, housekeeper seal; *D*, soft-soldered joint; *E*, rubber sleeve (section of automobile inner tube); *F*, felt ring; *G*, steel tube; *H*, interspace pumping lead; *I*, electric contact to silver coating of inner Dewar (for shielding); *J*, helium return line; *K*, rubber O ring and flange connection; *L*, transfer-tube port and O-ring seal; *M*, high-voltage lead-in; *N*, eight-wire lead-in (only two shown); *O*, bakelite supports; *P*, unsilvered area of inner Dewar; *Q*, Kovar–glass seal; *R*, liquid-nitrogen inlet (exit not shown).

Photography

With ordinary screens the intensity available with currents of 10^{-7}–10^{-6} amp is so ample that ordinary optics and films are adequate. It is even possible to take 16-mm motion-picture films at exposures of 1/30 sec per frame with $f/1.4$ lenses and fast commercial films.[18] The intensity available in ion microscopy is much less. Hydrogen-ion still pictures can be taken with fast films [18] and $f/1.4$ optics, but reasonable exposure times with helium-ion images require $f/1.0$ lenses. A number of these are commercially available, usually at high prices.[19] A number of $f/1.0$ lenses for 16-mm film are available [20] in the $100–$200 range.

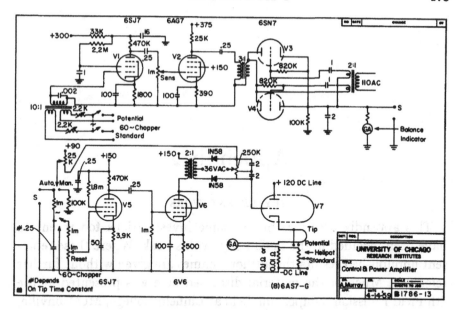

Fig. 129. Circuit diagram of tip-temperature controller: double chopper, James Instrument Chopper Model C-1275 or Stevens-Arnold Inc. Model A-11-12 dc-ac chopper; single chopper, dc millisec relay type 152; V_7 is a bank of up to eight 6AS7-G tubes in parallel, the number being controlled by an external manual switch. These tubes are operated from a specially regulated ripple-free supply or batteries, the remainder from a well-regulated electronic supply. The regulator is of the sampling type and amplifies the difference between the voltage generated in the standard resistor and that developed between the tip's potential leads. This ac signal is rectified in a phase detector and actuates the final power amplifier, which is arranged for inverse feedback. See also Fig. 61.

Appendix 2

VERY RECENT WORK

This appendix describes briefly some investigations too recent to have been included in the main body of this book. Where these represent unpublished results the authors' names are given without further reference. Much of the material discussed here was presented at the 7th Field Emission Symposium, held at Linfield College, McMinnville, Oregon, August 31–September 2, 1960.

Field-Ion Microscopy

E. W. Müller has succeeded in increasing the brightness of He⁺ ion images by a factor of 100 without appreciable loss in resolution. The method consists in using a very small emitter-to-cathode spacing, so that helium pressures up to 80×10^{-3} mm-of-mercury can be used without causing discharges. The ion beam reaches the screen through a small hole (1-mm diameter) in the cathode, which is equipotential with the screen. The space between cathode and screen is pumped separately and is at low pressure, to avoid defocusing collisions. Since high voltages are used for image formation, the screen life is relatively short.

Müller has also found that H_2O and N_2 attack tungsten at He⁺-ion-forming fields. The effect does not seem to be sputtering but is electrochemical in nature. Details are still obscure, but the difficulty may be avoided by using purified helium.

Müller has investigated cathode sputtering, bombardment by 10-kev neutral He atoms, and α-bombardment of tungsten and platinum emitters. By printing successive micrograms on red and green film, it is possible to detect all changes that have occurred; when the films are superimposed deletions from the original pattern show up red, additions green, and the unchanged portions yellow. Müller could show that sputtering caused severe damage to the underlying material as well as surface pile-ups. In the case of α-irradiation, damage was observed only on the exit side in the form of displacement spikes. Work is under way to correlate the direction of these spikes with that of the α-particles. By peeling off atom layers the presence of vacancies and interstitials resulting from α-passage through the emitter could be demonstrated. Somewhat similar results were obtained after bombardment with neutral atoms, produced by charge exchange.

G. Ehrlich has examined tungsten emitters after the adsorption of nitrogen, which is not field desorbed at He^+-ion-forming fields. He finds that individual N atoms are visible as somewhat larger spots than tungsten atoms, and is thus able to determine the preferred sites for adsorption or dissociation. These turn out to be the regions where ad-atoms can interact with the largest number of nearest-neighbor substrate atoms.

Work-Function Measurements

R. D. Young and E. W. Müller have succeeded in measuring work functions of individual planes on field-desorbed and hence almost ideally smooth tungsten emitters. A value of $\phi = 5.96$ ev for the 110 plane was obtained. Other regions will be investigated. Since the surface structure is known in detail from ion micrograms, it should be possible, for example, to correlate work function and emission with step density on the 110–211 zone which consists entirely of slabs of 110 orientation.

T. H. George and P. M. Stier have adsorbed oxygen on field-desorbed tungsten surfaces and find no marked work-function changes from thermally annealed substrates.

Field Emission from Semiconductors

R. L. Perry of the Linfield group has obtained current-voltage characteristics on conventionally prepared and thermally cleaned silicon emitters, whose dimensions were obtained by "ordinary" electron microscopy. He finds linear plots of log i as a function of $1/V$ for those cases where there is reason to believe that a sufficient number of surface states exists to screen out the field. In other cases the plot shows two linear portions separated by a curved one, in qualitative agreement with Stratton's theory (see Chapter 1).

F. G. Allen has succeeded in preparing and field desorbing germanium tips. He finds a very simple pattern, showing mobility at 300° K, after field desorption. After thermal annealing this reverts to a more complicated form which closely resembles that of silicon emitters obtained by him and by the Linfield group. It is not completely clear at the moment whether this means that the thermally annealed form represents a contaminated surface. The fact that it is identical with the patterns obtained by A. J. Melmed from whiskers grown *in situ*, which presumably were clean, suggests that clean annealed germanium surfaces may have a more complicated structure or work-function anisotropy than most metals.

Field Emission from Whiskers

A. J. Melmed and the author have followed the kinetics of gold whisker growth electrically (see Chapter 5). It could be shown that growth is positively exponential, as in the case of mercury, and that this is followed by thickening beyond a critical, diffusion-limited, length. It was possible to determine the latter as a function of substrate temperature and so to obtain a value of $E_{diff} = 24 \pm 5$ kcal for the surface diffusion of gold on gold.

Surface Diffusion

G. Ehrlich and F. G. Hudda [1] have studied the diffusion of nitrogen on tungsten and find all three types of diffusion (see Chapter 4). Spreading occurs with a sharp boundary near 40° K, with an estimated activation energy of < 1.9 kcal. Boundary diffusion in the chemisorbed layer occurs between 425° and 650° K, the boundary converging on the 100 poles as in the case of carbon monoxide. The activation energy for this process was estimated to be 20 kcal. Diffusion without a boundary over 116 and 120 occurs with an estimated activation energy of ~ 35 kcal. The ratio $E_{diff}/E_{des} = 0.23$ for this process, which agrees rather closely with the value for oxygen on tungsten. Since the atoms are of comparable size, this is a reasonable result.

B. Barnaby and E. A. Coomes have studied the diffusion of strontium and of strontium oxide on tungsten. In the former case boundary migration, analogous to the low-temperature spreading of chemisorbed gases, occurs near 450° K. The second layer is desorbed at 600° K, and the first one at 1200° K. In the case of strontium oxide there is evidence for dissociation and separate migration of oxygen and strontium. Desorption occurs via a complex series of steps, involving reaction with the substrate.[2]

Adsorption of Carbon Monoxide on Tungsten

L. W. Swanson and the author have measured the heat of adsorption of carbon monoxide on tungsten as a function of coverage and find it to increase from < 10 kcal at high to > 90 kcal at low coverage (Fig. 130). Heating of fully covered emitters led to irreversible changes in the sense that the work function and pattern resulting from low-temperature spreading over clean tungsten could not be approximated by redosing partially covered emitters. Instead, decreases in the work function were observed on redosing (Fig. 131).

The following mechanism has been provisionally postulated. When an initially clean tungsten emitter is covered with carbon monoxide by low-temperature spreading, nonactivated chemisorption occurs. When the surface is heated, desorption takes place, the amount

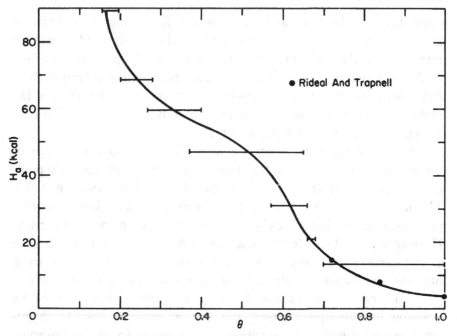

Fig. 130. Heat of adsorption of carbon monoxide on tungsten as a function of relative coverage θ. Lines indicate coverage intervals of desorption. Solid points after Rideal and Trapnell, *Proc. Roy. Soc. 205A*, 409 (1951).

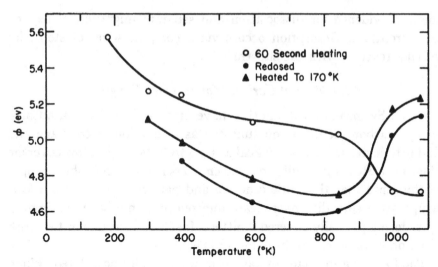

Fig. 131. Work function of tungsten emitters covered with carbon monoxide: ○ tip covered by low-temperature spread over clean tungsten, then heated 60 sec to indicated temperature; ● redosed with carbon monoxide after heating to point indicated by corresponding open circle, but not heated; ▲ tip treated as in ● but heated to 170°K.

depending on the time and temperature of heating. Desorption is accompanied by some rearrangement of the residual adsorbate. A number of molecules equal to that evaporated "flips over" and occupies two erstwhile sites. The new mode presumably corresponds to flat adsorption; in any case it probably involves some distortion of the carbon-oxygen bond and thus requires an activation energy. This may account for its nonoccurrence at 20° K.

Subsequent exposure to carbon monoxide leads to a new type of adsorption, different from the low-temperature and the "flipped" modes, since none of the original sites are available when $\theta \geqslant 0.5$. The new mode corresponds to fairly strong binding, but resembles physisorption in that it leads to a reduction in the work function. Heating a tip redosed in this manner leads to a gradual evaporation of the pseudo-physically adsorbed gas and eventually to the work function and pattern resulting from the initial heat treatment. When the latter leads to values of $\theta < 0.5$, some of the original sites will be available for subsequent adsorption, so that redosing leads to chemisorption and pseudo physisorption. It is possible to fit the observed data shown in Fig. 131, including the position of the crossover, with these assumptions.

Field Desorption of Water

H. D. Beckey has found that $H_3O^+ \cdot (H_2O)_n$ ions are produced from condensed water layers at very low fields (10^7 v/cm) and concludes that a high degree of a priori ionization exists in the film at fields of this magnitude. The observed process then corresponds to desorption of the ions already present, that is, to type 2 field desorption (see Chapter 3).

Field Desorption of Carbon Monoxide from Tungsten

L. W. Swanson and the author have used the field desorption of carbon monoxide from tungsten to obtain a potential curve for the adsorbed state by the method outlined at the end of Chapter 3. This curve is shown in Fig. 132. The rate constant for field desorption was found to have the form of Eq. (3.61) but contrary to initial expectations the transmission coefficient s was found to decrease steadily with decreasing activation energy of field desorption, even though this corresponded to successively higher fields. At $F = 2.88$ v/A desorption occurred with zero activation energy even at 20° K, but with $k_{des} = 10^{-2}$ sec^{-1}, which is extremely small compared to $1/\nu_{vib}$. However a 4 percent increase in F at this point increased the rate constant by 200.

This behavior suggests that the rate-controlling step at high fields and low temperatures is not the electronic transition, but tunneling of CO molecules *through* the potential barrier (Fig. 133). This effect is given added importance by the discrete energy values of the vibrational levels and should become most pronounced when their spacing is large compared to kT. Under these conditions the lowest level ($n = 9$, say) below the transition point may lie considerably (relative to kT) below it, so that tunneling provides a greater net rate of desorption than activation to the saddle point. This argument indicates that the heights of the tunneling barrier are $\geqslant 0.13$–0.05 ev in the present case.

If desorption occurs mainly by tunneling, the rate constant is, very roughly,

$$k_{des} = \nu e^{-(E_n - E_0)/kT} \exp\left[-2(2m_{CO}/\hbar^2)^{\frac{1}{2}} \int_{x_1}^{x_2} (V - E_n)^{\frac{1}{2}} dx\right], \quad (1)$$

where E_n is the energy of the nth vibrational level and E_0 that of the

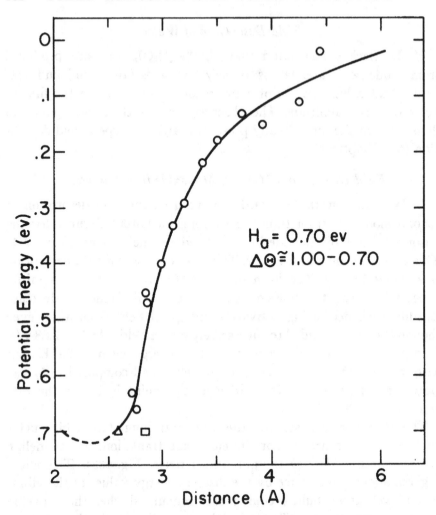

Fig. 132. Potential energy as a function of distance from the surface of electro-neutrality for carbon monoxide adsorbed on tungsten. Zero-field heat of adsorption, 0.70 ev. The coverage interval of the experiment is indicated on the figure. □ Zero desorption energy for $F = 2.88$ v/A. △ Zero desorption energy for $F = 3.00$ v/A.

ground state; and n corresponds to the highest level still under the saddle point of the barrier. In general, increases in F will not cause increases in the temperature-independent part of the rate constant, since the first-order effect is a reduction in n, that is in the energy of the level from which tunneling principally occurs. However, at sufficiently low temperature where only the ground state is occupied, or at fields so high that it is the only level below the saddle point,

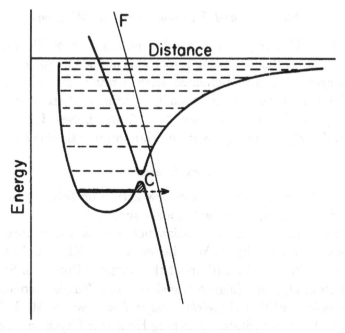

Fig. 133. Schematic diagram illustrating field desorption by tunneling (arrow) of the adsorbate through the potential barrier. Tunneling occurs from the highest vibrational level (here $n = 0$) still below the barrier. This level is somewhat broadened. In the diagram the field F is taken so high that all other vibrational levels would lie above the transition point C and are virtual only (dashed horizontal lines).

increases in F *must* decrease the tunneling barrier, and so lead to the observed behavior. Equation (1) indicates that the magnitude of barrier penetration is adequate to account for the observed values of k_{des}.

At "low" values of F the oscillator behaves more classically since the tunneling barrier is larger, the levels are more closely spaced, and the desorption temperature is higher, so that the temperature-independent part of the rate constant should increase, as is found experimentally.

The occurrence of tunneling requires corrections in the potential curve, which amount to a slight leftward shift of its lower part. These corrections have not been made in Fig. 132.

The experiment described here is interesting also because it gives strong evidence for the tunneling of massive particles.

Electronic and Electron-Optical Applications

The Linfield group has succeeded in building oscillographs with field-emission cathodes having resolutions of 2000 lines/in.

This group has also perfected flash x-ray tubes using a grid of heated 0.01-in. tungsten wires as a T–F cathode. Present designs yield 2000-amp, 600-kv electron beams for 0.2-μsec pulses. It is possible to make radiographs through 8 in. of aluminum with these techniques.

New Literature

E. W. Müller has written a review article on field-ion microscopy and field ionization, which will appear shortly.[3]

A survey, in English, of Russian field-emission literature through 1959 has been made by T. W. Marton and R. Klein and can be obtained by writing to the authors at the National Bureau of Standards, Washington, D.C. An English translation of a Russian monograph on field emission entitled *Avtoelektronnaya Emissiya* by M. I. Yelinson and G. F. Vasil'yev (State Publishing House for Physicomathematical Literature, Moscow, 1958) has been privately circulated but is not generally available as far as I know. This monograph treats the theory of field emission in considerable detail, with some emphasis on semiconductors. Field ionization is not included.

References

CHAPTER 1. THEORY OF FIELD EMISSION

1. (a) Summaries of work in this field through 1955 are contained in a review paper by R. H. Good and E. W. Müller, *Handbuch der Physik*, vol. 21 (1956), p. 176; (b) an earlier review paper, containing some original information not covered in (a), is E. W. Müller, *Ergeb. exakt. Naturwiss. 27*, 290 (1953).

2. The recognition of the importance of tunneling, or field emission, is due to Gamow, who explained α-emission by this mechanism. The application to the cold emission of electrons from metals was made by R. H. Fowler and L. W. Nordheim, *Proc. Roy. Soc. (London) A 119*, 173 (1928).

3. Lucid elementary discussions are given by F. Seitz, *Modern Theory of Solids* (McGraw-Hill, New York, 1940), chap. 4, and by C. Kittel, *Introduction to Solid State Physics* (Wiley, New York, 1953), chaps. 12–14.

4. An excellent summary of available information and theory is given by C. Herring and M. H. Nichols, *Revs. Modern Phys. 21*, 185–270 (1949).

5. H. Juretschke, in R. Gomer and C. S. Smith, eds., *Structure and Properties of Solid Surfaces* (University of Chicago Press, Chicago, 1953), p. 100.

6. (a) R. Smoluchowski, *Phys. Rev. 60*, 661 (1941); (b) R. Gomer, *J. Chem. Phys. 21*, 1869 (1953).

7. J. Mayer and M. Mayer, *Statistical Mechanics*, (Wiley, New York, 1940), p. 390.

8. A. L. Hughes and L. A. DuBridge, *Photoelectric Phenomena* (McGraw-Hill, New York, 1932).

9. See, for instance, V. Rojansky, *Introductory Quantum Mechanics* (Prentice Hall, New York, 1938).

10. W. Schottky, *Z. tech. Physik 14*, 63 (1923).

11. L. W. Nordheim, *Proc. Roy. Soc. (London) A 121*, 628 (1928).

12. R. D. Young, *Phys. Rev. 113*, 110 (1959).

13. R. D. Young and E. W. Müller, *Phys. Rev. 113*, 115 (1959).

14. W. P. Dyke and W. W. Dolan, in L. Marton, ed., *Advances in Electronics and Electron Physics*, 8 (Academic Press, New York, 1956), p. 89.

15. F. G. Allen, *J. Appl. Phys. 28*, 1510 (1957).

16. L. A. D'Asaro, *J. Appl. Phys. 29*, 33 (1958).

17. W. P. Dyke, private communication.

18. A. J. Melmed and R. Gomer, unpublished.
19. N. Margulis, *J. Phys. (U.S.S.R.)* **11**, 67 (1947).
20. R. Stratton, *Proc. Phys. Soc. (London)* B *68*, 746 (1955).

Review Articles

A number of review articles have appeared on field emission and related topics. Some of these are given above. They are retabulated here, for greater convenience, with other reference material.

Related Topics

Electron theory of metals: C. Kittel, *Introduction to Solid State Physics* (Wiley, New York, 1953).

Thermionic emission: C. Herring and M. H. Nichols, *Revs. Modern Phys.* **21**, 185 (1949).

Electronic properties of surfaces: H. Juretschke in R. Gomer and C. S. Smith, eds., *Structure and Properties of Solid Surfaces* (University of Chicago Press, Chicago, 1953), p. 100.

Mainly Field Emission

R. O. Jenkins, *Repts. Progr. Phys.* **9**, 177 (1943).

F. Ashworth, in L. Marton, ed., *Advances in Electronics*, 3 (Academic Press, New York, 1951), p. 1.

E. W. Müller, *Ergeb. exakt. Naturwiss.* **27**, 290 (1953).

R. Gomer, *Advances in Catalysis*, 7 (1955), p. 93.

W. P. Dyke and W. W. Dolan, in L. Marton, ed., *Advances in Electronics and Electron Physics*, 8 (1956), p. 89.

R. H. Good and E. W. Müller, *Handbuch der Physik* (Springer, Berlin, 1956), vol. 21, p. 176.

CHAPTER 2. FIELD-EMISSION MICROSCOPY AND RELATED TOPICS

1. Reference 1, Chapter 1.
2. D. J. Rose, *J. Appl. Phys.* **27**, 215 (1956).
3. A. J. Melmed and E. W. Müller, *J. Chem. Phys.* **29**, 1037 (1958).
4. R. Gomer and D. A. Speer, *J. Chem. Phys.* **21**, 73 (1953).
5. R. Gomer, *J. Chem. Phys.* **21**, 293 (1953).
6. The problem of resolution is discussed *inter alia* by (*a*) G. Richter, *Z. Physik* **119**, 406 (1942); (*b*) E. W. Müller, *Z. Physik* **120**, 270 (1943); (*c*) R. Gomer, *J. Chem. Phys.* **20**, 1772 (1952). See also references 1 and 7.
7. W. P. Dyke and W. W. Dolan, in L. Marton, ed., *Advances in Electronics and Electron Physics*, 8 (Academic Press, New York, 1956), p. 89.
8. C. Herring, *Phys. Rev.* **82**, 87 (1951).
9. F. M. Charbonnier, private communication.
10. This equation is also given by R. Klein, *J. Chem. Phys.* **31**, 1306 (1959).

11. This effect was first discussed by J. A. Becker, private communication.
12. R. Gomer, *J. Chem. Phys. 21*, 1869 (1953).
13. M. Drechsler, 1959 Field Emission Symposium.
14. R. Gomer, *J. Chem. Phys. 28*, 458 (1958).
15. A. J. Melmed and R. Gomer, *J. Chem. Phys. 30*, 586 (1959).
16. E. C. Cooper and E. W. Müller, *Rev. Sci. Inst. 29*, 309 (1958).
17. The following discussion is a slight expansion of arguments given by C. Herring in R. Gomer and C. S. Smith, eds., *Structure and Properties of Solid Surfaces* (University of Chicago Press, Chicago, 1953), p. 63. Equation (55) of the present work is essentially that given also by Dyke and Dolan, reference 7.
18. E. W. Müller, *J. Appl. Phys. 26*, 732 (1955).
19. P. C. Bettler, unpublished.
20. I. L. Sokolskaya, *J. Tech. Phys. (U.S.S.R.) 26*, 1177 (1956).
21. M. Drechsler, private communication.
22. R. Klein, *J. Chem. Phys. 21*, 1177 (1953).
23. R. Gomer and J. K. Hulm, *J. Chem. Phys. 27*, 1363 (1957).
24. E. W. Müller, *Ergeb. exakt. Naturwiss. 27*, 290 (1953).
25. C. Herring and J. K. Galt, *Phys. Rev. 85*, 1060 (1952).
26. L. Marton and R. A. Schrack, unpublished.

Review Articles

Articles dealing specifically with field emission are listed at the end of Chapter 1. The following are pertinent to various applications:

Thermodynamics of surfaces and related topics: C. Herring in Gomer and Smith, *Structure and Properties of Solid Surfaces*, p. 5.

Surface phenomena: R. Gomer, in J. E. Goldman, *Science of Engineering Materials* (Wiley, New York, 1957), p. 216; Report of Committee on Perspectives in Materials Research, National Academy of Sciences, to be published (this is a 190-page summary of most phases of surface physics).

Chemisorption. See references to Chapter 4.

Chapter 3. Field Ionization and Related Phenomena

1. M. G. Inghram and R. Gomer, *Z. Naturforsch. 10a*, 863 (1955).
2. (a) R. H. Good and E. W. Müller, *Handbuch der Physik* (Springer, Berlin, 1956), vol. 21, p. 176; (b) E. W. Müller, *J. Appl. Phys. 28*, 1 (1957); (c) E. W. Müller, *Z. Physik 156*, 399 (1959).
3. E. W. Müller, *Phys. Rev. 102*, 618 (1956).
4. R. Gomer, *J. Chem. Phys. 31*, 341 (1959).
5. E. W. Müller, *Ergeb. exakt. Naturwiss. 27*, 290 (1953).
6. R. Gomer, *J. Chem. Phys. 20*, 1772 (1952).
7. P. Schissel, private communication.
8. H. D. Beckey, *Z. Naturforsch. 14a*, 712 (1959).

9. S. Glasstone, K. Laidler, and H. Eyring, *The Theory of Rate Processes* (McGraw-Hill, New York, 1943).

10. I. Langmuir, *J. Franklin Inst. 217*, 543 (1934).

11. E. W. Müller, *J. Appl. Phys. 26*, 732 (1955).

12. M. G. Inghram and R. Gomer, unpublished.

13. G. Ehrlich, *J. Phys. Chem. 60*, 1388 (1956).

14. E. C. Cooper and E. W. Müller, *Rev. Sci. Inst. 29*, 309 (1958).

CHAPTER 4. SOME APPLICATIONS OF FIELD EMISSION TO ADSORPTION

1. I. Langmuir, *Trans. Faraday Soc. 17*, 1 (1921).

2. S. Brunauer, P. H. Emmett, and E. Teller, *J. Am. Chem. Soc. 60*, 309 (1938).

3. G. Halsey, *J. Chem. Phys. 16*, 931 (1948).

4. R. Gomer, *J. Chem. Phys. 29*, 441 (1958); *J. Phys. Chem. 63*, 468 (1959).

5. T. L. Hill, in R. Gomer and C. S. Smith, eds., *Structure and Properties of Solid Surfaces* (University of Chicago Press, Chicago, 1953).

6. G. Ehrlich and F. G. Hudda, *J. Chem. Phys. 30*, 493 (1959).

7. J. C. P. Mignolet, *J. Chem. Phys. 23*, 753 (1955); *Discussions Faraday Soc.*, 8 (1950), p. 105; *Rec. trav. chim. 74*, 701 (1955).

8. J. I. Roberts, *Proc. Roy. Soc. A 152*, 445 (1935); *161*, 141 (1937).

9. A. Wheeler, in Gomer and Smith, *Structure and Properties of Solid Surfaces*.

10. (a) R. Gomer, R. Wortman, and R. Lundy, *J. Chem. Phys. 26*, 1147 (1957); (b) R. Wortman, R. Gomer, and R. Lundy, *J. Chem. Phys. 27*, 1099 (1957); (c) R. Gomer and J. K. Hulm, *J. Chem. Phys. 27*, 1363 (1957).

11. J. A. Becker, *Advances in Catalysis*, 7 (Academic Press, New York, 1955), p. 135.

12. G. Ehrlich, *J. Phys. Chem. 60*, 1388 (1956); P. Kisliuk, *J. Phys. Chem. Solids 3*, 95 (1957).

13. G. Ehrlich, T. W. Hickmott, and F. G. Hudda, *J. Chem. Phys. 28*, 506 (1958).

14. D. D. Ely, *Discussions Faraday Soc.*, 8 (1950), p. 34; D. P. Stevenson, *J. Chem. Phys. 23*, 203 (1955).

15. P. W. Selwood, *Advances in Catalysis*, 9 (1957), p. 93.

16. R. Suhrmann, *Advances in Catalysis*, 7 (1955), p. 303.

17. K. Huang and G. Wyllie, *Discussions Faraday Soc.* 8 (1950), p. 18.

18. M. McD. Baker and G. I. Jenkins, *Advances in Catalysis*, 7 (1955), p. 1.

19. D. Alpert and R. Buritz, *J. Appl. Phys. 25*, 202 (1954) and previous papers.

20. R. Gomer, *J. Chem. Phys. 28*, 168 (1958) and unpublished results.

21. D. O. Hayward and R. Gomer, *J. Chem. Phys. 30*, 1617 (1959).

22. T. W. Hickmott, *J. Chem. Phys. 32*, 810 (1960); *Discussions Faraday Soc.*, 28 (1959), p. 60.

23. E. W. Müller, private communication.

24. J. A. Becker and R. G. Brandes, *J. Chem. Phys. 23*, 1323 (1955).

25. R. Klein, *J. Chem. Phys. 31*, 1306 (1959).

Review Articles on Adsorption

A. Wheeler, in Gomer and Smith, *Structure and Properties of Solid Surfaces*, p. 439.

G. Ehrlich, *J. Phys. Chem. Solids 1*, 1 (1956).

K. Hauffe, *Advances in Catalysis*, 7 (1955), p. 213.

M. McD. Baker and G. I. Jenkins, *Advances in Catalysis*, 7 (1955), p. 1.

CHAPTER 5. SOME MISCELLANEOUS APPLICATIONS OF FIELD
EMISSION

1. (a) R. Gomer, *J. Chem. Phys. 29*, 443 (1958); (b) R. Gomer, *J. Austral. Phys. 13*, 391 (1960).

2. E. Fermi, *Nuovo Cimento 11*, 157 (1934).

3. W. P. Allis and P. M. Morse, *Z. Physik 70*, 567 (1931).

4. For references to this and earlier work, see A. J. Melmed and E. W. Müller, *J. Chem. Phys. 29*, 1037 (1958).

5. A. J. Melmed, private communication.

6. R. Gomer and D. A. Speer, *J. Chem. Phys. 21*, 73 (1953).

7. R. H. Good and E. W. Müller, *Handbuch der Physik* (Springer, Berlin, 1956), vol. 21, p. 176.

8. G. W. Sears, *Acta Met. 3*, 361 (1955).

9. (a) R. Gomer, *J. Chem. Phys. 28*, 457 (1958); (b) R. Gomer, in R. H. Doremus, ed., *Growth and Perfection of Crystals* (Wiley, New York, 1958), p. 126.

10. J. D. Eshelby, *J. Appl. Phys. 24*, 176 (1953).

11. J. D. Eshelby, in Doremus, *Growth and Perfection of Crystals*, p. 130.

12. A. J. Melmed and R. Gomer, *J. Chem. Phys. 30*, 586 (1959).

13. A. J. Melmed and D. O. Hayward, *J. Chem. Phys. 31*, 545 (1959).

APPENDIX 1. EXPERIMENTAL ASPECTS

1. E. W. Müller, in *Physical Methods of Chemical Analysis, III*, Academic Press, New York (1956).

2. R. Gomer, *Advances in Catalysis*, 7 (Academic Press, New York, 1955), p. 93.

3. When the prevention of helium leakage through the walls is important special glasses or ceramics must be used. See W. P. Dyke and W. W. Dolan, in L. Marton, ed., *Advances in Electronics*, 8 (Academic Press, New York, 1956), p. 89.

4. These are available from a number of suppliers. I am happy to acknowledge here my indebtedness to Dr. Samuel Eisenberg and the Sampson Chemical and Pigment Corporation, 2830 West Lake Street, Chicago, Illinois for their generous aid in supplying phosphors, pertinent information, and instruction.

5. I first learned of this technique from Professor E. W. Coomes of Notre Dame University. The method has also been used by E. W. Müller.

6. R. Gomer, *Rev. Sci. Inst.* *24*, 993 (1953). This is a modification of a method used industrially, for instance under the name NESA by the Pittsburgh Plate Glass Company.

7. To the best of my knowledge this technique was originated in.the laboratories of the Linfield group under W. P. Dyke.

8. R. Gomer, *Rev. Sci. Inst.* *27*, 544 (1956).

9. E. W. Müller, *Ergeb. exakt. Naturwiss.* *27*, 290 (1953).

10. R. Gomer, R. Wortman, and R. Lundy, *J. Chem. Phys.* *26*, 1147 (1957).

11. D. Alpert and R. Buritz, *J. Appl. Phys.* *25*, 202 (1954).

12. Available from the Electronic Tube Division, P. O. Box 284, Elmyra, New York.

13. Consolidated Electrodynamics Corporation, Rochester Division, Rochester 3, New York.

14. A good description of cryogenic techniques can be found in R. B. Scott, *Cryogenic Engineering* (Van Nostrand, Princeton, 1959).

15. For instance, from the Beta Division of Sorensen and Company, Norwalk, Connecticut.

16. Manufactured by the Applied Physics Corporation, 2724 South Peck Road, Monrovia, California.

17. A very good battery-operated electrometer is manufactured by the Keithley Corporation, 12415 Euclid Avenue, Cleveland, Ohio.

18. Such as Tri-X, Royal X Pan, or Polaroid 3000.

19. Wray Ltd., Bromley, Kent, England.

20. The author has found the Carl Meyer 1-in. focal length $f/0.95$ *Moviar* very satisfactory.

Appendix 2. Very Recent Work

1. G. Ehrlich and F. G. Hudda, *J. Chem. Phys.* *32*, 924 (1960).
2. J. A. Cape and E. A. Coomes, *J. Chem. Phys.* *32*, 210 (1960).
3. E. W. Müller, *Adv. Electronics and Electron Physics 13*, 83 (1960).

Index

Printed in the United States
By Bookmasters